BETTER TECH

BETTER TECH

Putting *People* First in Cyberspace

FRANCINE BERMAN

The MIT Press
Cambridge, Massachusetts
London, England

The MIT Press
Massachusetts Institute of Technology
77 Massachusetts Avenue
Cambridge, MA 02139
mitpress.mit.edu

The MIT Press would like to thank the anonymous peer reviewers who provided comments on drafts of this book. The generous work of academic experts is essential for establishing the authority and quality of our publications. We acknowledge with gratitude the contributions of these otherwise uncredited readers.

This book was set in Scala and ScalaSans by New Best-set Typesetters Ltd. Printed and bound in the United States of America.

Library of Congress Cataloging-in-Publication Data is available.

ISBN: 978-0-262-05488-1

10 9 8 7 6 5 4 3 2 1

EU Authorised Representative: Easy Access System Europe, Mustamäe tee 50, 10621 Tallinn, Estonia | Email: gpsr.requests@easproject.com

"If you wish to move mountains tomorrow, you must start by lifting stones today."
—African proverb

CONTENTS

I

WHERE CYBERSPACE FAILS US

PROLOGUE, PART I

Technology is not a religion. But sometimes we treat it like one, worshiping powerful companies and their apps and services like gods who need to be appeased, residing in an uncontrollable cyberspace heaven. Thinking about technology this way creates an unhealthy power dynamic, one where tech solutions are right by default and our inability to thrive in the cyberspace ecosystem is our fault. Who is to blame when your problem doesn't exist in a company's automated customer service system, or when an AI "decision" puts you at a disadvantage, or when you're not represented in someone's user database? It's not you.

What if we engaged with technology with a different set of expectations? A set of expectations that puts technology in a public service role and puts people rather than companies first. What if we assumed that if technology puts us at risk, the risk should be lowered or mitigated instead of us just getting used to it? That if technology doesn't serve people well, the technology needs to be fixed, not the people. I wrote this book to partner with you in thinking about how we can do a better job of managing the technologies we depend on most and reframing cyberspace in the public interest. So you and I and others can nudge cyberspace to be more beneficial and less risky for the people who use it.

Technology that puts people first isn't just a nice-to-have. It's crucial for a present and future where society advances and humanity thrives. The information technologies we depend on have become critical infrastructure vital to the way we live. We pay our rent, taxes, and parking meters online. We use automated systems to apply to and process applications for schools, jobs, services, and loans. AI-powered systems drive cars, populate factories, and steer us toward (and away from) opportunities, products, and opinions. Digital systems are essential to fully participate in today's world. They have become as important for societal functioning and well-being as roads and bridges, food and health systems, and the electrical grid—all considered critical infrastructure needed to participate, function, and thrive.

But as digital systems and services become more essential to modern life, we continue to treat them as successful commercial products rather than critical infrastructure we depend on. We don't expect public interests to take precedence over business interests. We don't expect digital systems and products to be created, deployed, and managed like other critical infrastructure—to support the public and protect us from harm. We take what we get.

What if we began to expect the most essential information technologies—the ones we depend on to function and for our well-being—to meet the norms we apply to other critical infrastructure? What if we expected these digital systems to be safe and trustworthy? To be designed and deployed to empower and protect, rather than exploit, the public? To be fixed when they fail and mitigate harms when they occur?

Imagine that the information we depend on, and that our social media and search engines serve up, was as carefully vetted for quality and risk as our food. Food and agriculture systems are critical infrastructure. Scrutiny from the US Department of Agriculture, the Food and Drug Administration, and other agencies gives us confidence that the food we buy won't make us sick and meets reasonable standards for public health. What if the algorithms driving critical apps and services, the data they use, and their ranking methods were audited to evaluate the integrity and veracity of their outcomes? Suppose that assessment was available to you—roughly similar to knowing whether your favorite restaurant received an A or a C rating at their last health code evaluation, something you might want to consider when you choose where to eat. Imagine that there was as much transparency about where your information came from and how it was handled as there is about where your food comes from and how it was processed. Knowing the landscape gives you options—and confidence—about what you consume. Wouldn't it be great to feel that way about your information?

What if we expected that when digital critical infrastructure puts us at serious risk that the problems would be addressed so that harms are reduced and bad outcomes are minimized? Water systems are critical infrastructure. The last glass of water you drank was likely unremarkable. You turned on the tap, filled your glass, and trusted the many systems that deliver your water and make sure it is safe to drink: systems that move water from its source to its destination, systems that clean water through the addition of chemicals and the use of various methods for filtration and disinfection, systems that ensure that your water meets health standards and flows easily through your tap. We rely on those systems. And for the most part, we don't give them even a second thought. But that wasn't the case in Flint, Michigan in 2014 when the water system failed to deliver safe, healthy water and began to make people sick.

In 2014, as part of a cost-cutting move, officials in Flint shifted the city's water source from Lake Huron and the Detroit River to the Flint River. Lead seepage from corroded pipes resulted in foul-smelling, orange-tinged water that exposed residents (and particularly children) to dangerous levels of lead and possibly an outbreak of Legionnaires' disease. The mayor of Flint, the governor of Michigan, and the president of the United States all declared a state of emergency. Governor Rick Snyder offered a stark assessment: "Let me be blunt. This was a failure of government at all levels. Local, state and federal officials—we all failed the families of Flint."[1] Over the next months and years, scientists, civil servants, and agencies worked to understand and address the issues, assess responsibility, navigate political systems, and, most importantly, develop remediations. Critical infrastructure had failed. Those involved were expected to be held accountable and everyone expected the problems to be addressed.

Digital critical infrastructure can fail as well, sometimes in ways that cause humans harm. Algorithms and AI can put people at risk—through cybersecurity vulnerabilities in your connected pacemaker or by misclassification of a stop sign for a speed limit posting in your self-driving car. Through unintentional biases built into the systems used to process your application for a loan, decide your eligibility for insurance or services, or sentence you in a court of law. Developed as commercial products, technologies are often designed to engage or addict us, and to gather tremendous amounts of valuable personal information used for system functioning and the pursuit of market share. What happens when these technologies pollute the social environment with misinformation, encourage divisiveness and predators, limit our choices and opportunities, or facilitate emotional or physical harm? Often, not much. When technologies become so essential that they are considered digital critical infrastructure, shouldn't we expect the priorities to shift? To elevate public interests and not just business interests? To minimize and mitigate harm?

It's time for us to expect digital critical infrastructure to meet the same public interest norms as other kinds of critical infrastructure. To support us and protect us as we navigate the world. To be safe, fair, accessible, and trustworthy. To function well by default and be fixed when problems arise that put us at dangerous risk. That's digital critical infrastructure that will empower humanity, even as cyberspace expands in capability and impact. That's the digital critical infrastructure humanity deserves.

Changing our expectations about digital critical infrastructure is the first step. It must be followed by *action* that makes tech better and better for us. We must change the way we create, use, and manage essential technologies to better promote the public interest. Doing so is a collective responsibility—none of us can do this alone. Even the most insanely powerful, wealthy, and famous among us can't improve cyberspace in one fell swoop. It requires a public interest–focused mindset and the persistent application of pragmatic and responsible strategies from those who build, use, and manage technology—all of us who want to reside in a world where tech is more empowering and less risky.

We can all do something. Each of our efforts can mount up and, over time, change things. After all, humans made cyberspace. And we have the power to make it better. If each of us adopts a public interest–focused mindset, we can reap more of the benefits and reduce the risks of cyberspace. The time to do this is *now*. With the advancement of AI and its encroachment on human function and draw on natural resources, we must change our expectations about the technologies essential to our well-being. We must begin to act intentionally to build and manage technologies that better serve the public. This book will help us develop a public interest tech mindset and provide a broad set of strategies to inform our actions.

Let me introduce myself. I fell in love with math as a kid and have been in computer science, often on the frontlines, for my entire professional career. Like everyone else over 30, I grew up in a different world. There was no Facebook, Google, TikTok, or Amazon.com. Reliable information came from the six o'clock news or a set of encyclopedias—information that people trusted to be reasonably accurate. My community consisted of people I had met and who knew me, at least a little. My life was private, and I could have secrets. If I had a problem, I could easily talk with a human who could get to the bottom of things and help make systems work for me.

That world gradually ceased to exist with the rise of the internet's most powerful apps and services. The present world bears little resemblance to the one I grew up in. Today, information comes from everywhere but it's hard to know what's fake and what's real. My digital profile has become my "permanent record"—aggregating and assessing my interests for systems that want to "know" me, define me, and target me. Very little about me is secret and everything known about me is out of context and not always true. Algorithms use my digital profile to make decisions about me, and when there's a problem, it's often hard to find a human who can get to the bottom of things.

How did we get here?

Technological innovation has transformed the world we live in and how we engage with it, often in exciting ways that broaden our horizons and expose us to a more diverse and expansive landscape, and often in ways that put us at risk for exploitation, manipulation, and worse. Cyberspace—comprised of the internet, its devices, and its systems, apps, and services—provides infrastructure for work, commerce, connection, and life. Its most essential technologies are critical infrastructure—vital to the way we function and fundamental to full participation in today's workforce, communities, and human society.

There is a feedback loop between how cyberspace operates and how society functions. We saw that clearly during the COVID-19 pandemic: In 2020 when seemingly everything shut down, cyberspace made it possible for many of us to maintain some semblance of professional and social functionality and important human connections. During 2020, I taught my courses remotely, bought groceries on Instacart, and got everything else on Amazon. I went to a Zoom wedding and a Zoom funeral. I took ballet lessons online twice a week with a portable barre I ordered from the internet. Cyberspace reliably provided the infrastructure that I and many others needed when it wasn't possible or safe to do things in person. It was a lifesaver.

But while we were using tech, tech was using us. Society's expanded use of cyberspace during the pandemic increased the risks of digital technologies. It stoked the unfettered collection of massive amounts of personal data—data that is regularly used to manipulate or exploit us. It exacerbated the spread of misinformation—some of it touting COVID "treatments" that could kill you. It widened the digital divide—the gap between those with access to computers and internet bandwidth and those without—the haves and have-nots of cyberspace.

The pandemic exposed the rewards and risks of digital technologies, both of which continue to increase today with our use of technology as critical infrastructure. Many of the risks come from the lack of public protections baked into the design, development, and management of essential online apps and services—search engines, commercial systems, AI-driven facial recognition systems, social media. These risks arise because better privacy protections, more security testing, and user and cultural accommodations often make technology more expensive to develop and less profitable to deploy. Because developing a commercial product often means prioritizing profit and competitive advantage over privacy, fairness, security, sustainability, and other public interests.

To be fair, it's not surprising that essential cyberspace technologies weren't crafted to be public-focused critical infrastructure from the outset. Most of today's essential technologies of cyberspace were not essential when they were developed. Google was born as an innovative approach for searching the internet in the '90s but there were plenty of other ways to get information. Facebook was born as a way to connect university students. Amazon was born as an online bookstore when most people bought their books at local bookstores or the mall. These services were invented, developed, and nurtured by private sector companies to capture market share and make a profit. Over time, their scope, audience, impact, and profitability increased. It's one thing to create a new product. It's entirely another to expect that future generations of that product will be safe enough, secure enough, private enough, and fair enough for wide-scale public use when everyone comes to depend on it.

Between their humble beginnings as commercial products and today's widespread dependence on them as critical infrastructure, today's online search, e-commerce, social media, and a host of other tech systems and services have become an essential part of modern life—*digital critical infrastructure*. Like other critical infrastructure, it's important for them to be safe and trustworthy enough for broad public use. Companies agree with this, but there are few laws or regulations to serve as a forcing function to make their public interest intentions actionable, and to dissuade them from prioritizing profitable features over public protections. Rampant data exploitation, security breaches, and misinformation are now all part of the landscape of cyberspace. Without sufficient public protections or safe alternatives, essential parts of cyberspace have become untrustworthy, predatory, and out of (social) control. When the risks of digital critical infrastructure threaten to overwhelm its rewards, how can society thrive?

The most essential parts of cyberspace should be subject to the same public scrutiny and societal controls as other forms of critical infrastructure. But today's most essential digital technologies mostly get a free pass. There are more rules, regulations, policies, and government agencies involved in ensuring that the apple you buy at your local grocery store won't make you sick than there are regulations and policies that ensure that the software you download on your computer will keep your data private, secure, and safe from exploitation.

Going forward, we should better manage the riskiest and most essential parts of cyberspace through responsible design, development, deployment, use, and oversight. We've used this approach successfully to manage other kinds of critical infrastructure: We design nuclear power plants with redundant safety and monitoring systems so we can better deal with problems before they happen. Our medicines go through independent vetting processes to make sure they are good for us. Federal law minimizes personal risk by putting your financial services company on the hook—not you—when your credit card is stolen and fraudulent charges mount up.

None of the strategies we use to lower the risks of critical infrastructure are perfect, nor is the polarized public sector where many of them are initiated. Risk reduction strategies reduce harm but rarely solve problems completely. But perfection should not be the enemy of good. For the most part, we can rely on critical infrastructure to work: Our cars are safer, our planes rarely fall from the sky, and our banks are (mostly) solvent. Responsible design and deployment, public scrutiny, and proactive interventions can, and do, help make critical infrastructure safer and more empowering. We should use similar approaches to mitigate the risks of the essential technologies of cyberspace.

The many stakeholders of cyberspace—users, companies, governments, organizations—can *all* do something to promote public benefits and reduce public harm. Many efforts can nudge cyberspace to be safer, more secure, more private, more ethical, more fair, less discriminatory, and more protective of the public good.

Needless to say, this isn't easy, both in this moment and for an ecosystem as pervasive and complex as cyberspace. It seems especially difficult when public protections are scarce or under attack. But that's exactly when our efforts are needed most. The real, sustained, and positive change we seek will not come from one "magic bullet" law, policy, or action. Instead, multiple efforts will add up and, over time, begin to define something in aggregate, like the tiles in a mosaic. With persistent and widespread action, our efforts to promote better tech can begin to make a difference, even if momentum is slowed by the shifting dynamics of politics and power. Our actions and changing expectations can turn the tide. Things will begin to improve. There's reason for hope.

Our digital critical infrastructure is human-made. And it is too dangerous. Without intervention and management, our essential technologies will likely become more so in the future, especially with today's increasingly powerful applications of AI. But we created these technologies, and we can design, build, and manage them better. Our well-being depends on it. You and I can do something—no matter what our jobs or roles in society—so that the technologies we depend on are better, and better for us. This book can help us think about how.

1

THE BIG BANG OF TECH

When my wallet was stolen in the train station in New York City, it was a dispiriting reminder of how much I depend on cyberspace and the limitations of the digital systems I rely on. If I had lost my wallet 20 or even 10 years ago, the thief would have taken only my wallet and my money. That would have been discouraging and inconvenient. Now it is also dangerous, as the thief took my identity, wreaking havoc as virtual me.

On the train home, I realized that my wallet was gone when my smartphone alerted me to $3,000 worth of purchases at Nordstrom's and Macy's, purchases I had not made. Gone were my credit cards, my driver's license, money, insurance cards, my COVID card (remember those?), and various other forms of ID and documentation.

I called the fraud number for my credit card company and talked with actual humans who could shut down my credit cards immediately. The people I talked with were great, sympathetic, and expeditious, but federal law makes my credit union the victim in this case—the inconvenience is mine, but the loss is theirs. I was lucky that there was no parking gate needing my credit card at the train station at home, or I wouldn't have been able to get my car out of the lot.

The ensuing days were spent in customer service hell as I tried to resurrect documentation of my existence and get new credit cards. My local bank gave me cash which I used to pay for gas and groceries, but I couldn't do anything that involved credit—parking, automatic bill pay, rent, online orders—making it critical to replace the cards quickly.

My credit union sent a new credit card via FedEx. FedEx had problems with my address and delayed the delivery. I couldn't find a local phone number for any of their facilities or actual humans who could help fix the problem. All phone numbers led to a computerized system that was so limited it just hung up on me when I exhausted its algorithmic options—I was a leaf that didn't exist in an automated decision tree.

It added insult to injury when a week after the theft, the thief went into a different branch of my bank and withdrew another $1,200 using a doctored version of my driver's license and my social security number. My social security number and bank account number were not in my wallet but were readily available on the dark web, courtesy of a data breach of one of the many companies I've dealt with. This information was neither hard to find nor expensive—the going rate for a social security number on the dark web is less than $4.

It could have been worse. Digital identity theft has allowed thieves to steal tax refunds, open new accounts and loans in people's names, take over victims' email and social media, commit crimes in victims' names, and more. But data breaches, identity theft, exploitation, and deepfakes are becoming a normal part of cyberspace. Welcome to the Information Age.

Our experience with technology doesn't have to be this way. Digital technologies can become safer, more secure, and less risky. With effort and time (a lot of both), we can develop and manage the technologies we depend on to be more protective and empowering. There are many ways to do this. An emerging workforce of individuals is beginning to create and manage technology in a more socially responsible way—*public interest technologists.* With a broad perspective and strategies that can help put the public first, you can become, or may already be, one of them.

UMass professor Emily Nutwell describes public interest technologists as having a *mindset* that orients the way they approach problems. In particular, they operate with intentionality and the knowledge that *tech is human-made,* not neutral, and that their decisions have societal consequences. They *consider multiple public interests* and seek to *ensure that technology is beneficial for everyone,* not just specific groups. They *take an ethical approach* to their work and *incorporate open and transparent policies* in the creation and management of technology. Approaching problems with this mindset makes a difference; it creates fertile ground for action that nudges cyberspace to be more empowering and less risky for the humans who engage with it. This book will explore how we can find the public interest technologist in each of us and help make cyberspace a place where humanity can thrive.

LIVING IN CYBERSPACE: ECOSYSTEM FOR A TECH-POWERED WORLD

Today it is increasingly difficult—effectively impossible—to work, live, and enjoy full rights of citizenship and societal resources without engaging with digital technologies. Tech provides the infrastructure for information-gathering, human connection, entertainment, work, and education. If you use a smartphone, call a rideshare, order things online, post on social media, videoconference for work or school, or sign up for anything on a connected device, you are using tech and tech is using you. How information systems are structured, how your data is managed, who or what is making decisions, and what is happening behind the scenes impacts everyone.

Cyberspace is an ecosystem. It is dynamic, decentralized, and human-made. Many of its most common apps and services are essential to the functioning of modern society. Cyberspace has transformed the way we engage and who we are. And it's becoming increasingly risky for the humans who depend on it.

Depending on whom you ask, digital technologies will either usher society into a new world of tremendous opportunity or drive humanity to extinction. The truth is somewhere in between. We must find a way to minimize and mitigate technology's risks while reaping its rewards. Sometimes that seems impossible. Tech is created and largely controlled by the people who profit from it. For the rest of us, managing the risks of cyberspace and changing its trajectory seem increasingly out of reach—with obscene amounts of money at play, powerful entrepreneurs, and big multinational companies that have more influence than many countries. The tension between private interests (profit, market leadership) and public interests (privacy, safety, security, fairness) often creates unacceptable risk. What can the public do?

A lot of things. First, we have to recognize what the risks are. Information technologies today are born of innovation, and with new ideas we imagine the potential, but many of the perils may be unknown. We best understand the societal impacts of digital innovation when new ideas are released into the wild as commercial products for the public. Often vulnerabilities show up at scale, when enough people use a product and unanticipated risks begin to emerge. For example, social networks are of great value between trusted friends and colleagues. At scale, with a whole world of unknown posters, predators, and followers, social networks have become fertile environments for cyberbullying, the spread of misinformation, child exploitation, and other harms. We now need to do something proactive, whether it be curation of content, limitations on who can participate and how, ways to vet the trustworthiness of posts or posters, viable alternatives, or other strategies. The goal? Controls that allow us to effectively manage social networks to reduce their risks so we can better enjoy their benefits.

Once we understand the downsides of the technologies we use, it's up to us to act to make sure that our essential technologies are big on public benefits and low on risk. We must create societal controls and interventions—law, policy, standards, practices—that manage both deployment and use of existing technologies to mitigate their risks. We must proactively design and develop new technologies in a socially responsible way to minimize risk and incorporate public protections.

All of this takes time. Roughly 100 years ago, sociologist William Ogburn called the period of adjustment between innovation and the development of societal frameworks to manage their risks the "cultural lag."[1] He recognized this lag 100 years before the public introduction of ChatGPT and the frenzy about the need to manage AI and social media, and 100 years after child labor and environmental laws were created to manage new forms of mass production during the Industrial Revolution. The need for societal interventions to promote public well-being in the wake of powerful innovation is not a new phenomenon. Humanity has been here before.

And so it can happen with information technology. With mounting public concerns about the exploitation of data, identity theft, the risks of AI, the need for security in self-driving cars and connected medical devices, and a host of other public challenges, it is time to create the interventions that will minimize and mitigate the risks of technology for the people who use it. To do that, it's useful to understand how we got here.

THE INFORMATION BIG BANG: HOW CYBERSPACE STARTED

In the beginning, there was the internet. The internet—a network of networks—was seeded through many efforts in computer networking during the 1970s and 1980s, and particularly through the vision and funding from the US Department of Defense's Advanced Research Projects Agency (ARPA) and the National Science Foundation. Use of the internet as we now know it was made possible by Tim Berners-Lee, a physicist working at the CERN intergovernmental particle physics lab in Geneva in the 1980s, who invented the World Wide Web.

The World Wide Web was ground zero for modern internet applications. Berners-Lee invented the Web to be an information management system that could serve as a central repository for research results, techniques, and other information. Files could be put online to the server by researchers, and their colleagues could access the information remotely through a browser—an application that retrieves and displays information from websites. The driving "killer app" for the system was CERN's telephone directory. Berners-Lee invented an internet addressing system (URLs) and a system for networked communication between a website and a server (HTTP—hypertext transfer protocol)—establishing a model of communication used by over a billion websites on the internet today.[2]

Internet development and use expanded rapidly in the 1990s and beyond, defining modern global communication. Importantly, global information migrated from physical libraries, notes, newspapers, books, reports, and other sources to somewhere online. By 2007, paper-based storage constituted .0007 percent of our technological memory while 94 percent was in digital form.[3]

In a very real sense, the internet seeded an information big bang for human society. From the development of the internet's physical telecommunication networks and software protocols to manage information flow, to the subsequent galaxies of apps, services, systems, and products, the internet and its digital universe—cyberspace—have had far-reaching societal, economic, and environmental impacts and have transformed the modern human experience.

The notion of *cyberspace* recognizes that the internet and its digital galaxies are an ecosystem with their own sociotechnical culture. The term cyberspace was loosely defined in the 1980s by William Gibson in the sci-fi book *Neuromancer* as "a consensual hallucination experienced daily by billions of legitimate operators, in every nation, by children being taught mathematical concepts . . . A graphic representation of data abstracted from the banks of every computer in the human system. Unthinkable complexity."[4]

Today's many definitions of cyberspace are a little more down-to-earth, but most convey the sense of cyberspace as an integrated and interconnected environment. This is reflected in the characterization of cyberspace by none other than the Joint Chiefs of Staff of the US Department of Defense who consider cyberspace to be one of five interdependent domains alongside land, air, maritime, and space.[5] Like each of the Department of Defense's other interdependent domains, cyberspace—driven by digital creation and use—operates as an ecosystem with its own culture, ethos, and norms.

In its early years, cyberspace was interesting, useful, and nonessential. There were other ways, common ways, to do everything you could do on the internet—get information, communicate, collaborate. But by the early 2000s, the internet had become the delivery mechanism of choice for a wide variety of applications, applications Berners-Lee and its original developers never imagined, and "born digital" applications for which the internet was the *only* delivery mechanism.

Fast-forward to today. Our world is unimaginable without the internet and its apps and services, and it's the only world that anyone under 30 has ever known. Cyberspace didn't just change the way that information is stored and accessed, it changed the way we engage with information, and it changed society. It changed how we learn and what we learn, how we work and where we work, how we play and how we engage. It changed how we date, shop, drive, and express ourselves.

We have become residents of cyberspace. Most of us have many devices that connect to the internet, including our laptops, smartphones, e-readers, cars, smart doorbells, fitness trackers, baby monitors, connected medical devices, and so on. Those devices produce and process a tsunami of information—information that can greatly benefit us and information that feeds highly profitable business models for digital products, apps, and services with an insatiable appetite for data. For nearly half the population, this is just the way things have always been.

The internet seeded the technological big bang of cyberspace, but its extraordinary and ongoing expansion is all human-made. Humans make choices at every stage to create, commercialize, and manage the technologies of cyberspace. Each of those decisions has consequences that impact the societal and physical worlds we live in, and ultimately our well-being. For example, how an app or service deals with data defines its approach to privacy: Your dating site provider can decide whether to keep your sexual preference private or share it publicly or with commercial partners. The way your credit card company stores your data determines how vulnerable it is to data breaches and how secure your personal information is. These choices are made by humans. They define the benefits and risks of the services and apps that we use.

This gives us a lever for positive change—a way to make decisions differently in order to prioritize and protect the people who use technology. Part II of this book focuses on ways to apply a public interest tech mindset, and offers many pragmatic strategies that people, organizations, and governments can use to help make cyberspace a less risky place without substantially deterring the breathtaking pace of innovation. Part III explores how we can use this mindset to reframe our future.

THE CYBERSPACE ECOSYSTEM: HOW HUMAN PRIORITIES INFLUENCE TECH OUTCOMES

All tech apps and services are the result of a complex interplay of big and little human choices—about functionality and design, profit and market forces, material properties and availability, politics and tariffs, impact, cost, and more. Design, implementation, deployment, use, and management of technologies all involve these decisions. How we make them transforms innovation into products and products into outcomes. Technology is not created to be neutral. Most tech products have been created in the commercial sector to meet business goals—profitability and market leadership. Tech business models drive both creation and use.

The business model for many of the essential tech apps and services we use is to make these products essentially "free" to the user, and it is commonly (and truly) noted that if you're not paying for the product, *you* are the product. In exchange for providing apps and services to customers for free or minimal cost, companies vacuum up massive amounts of user data—data that has tremendous business and financial value to them. Many companies have data policies—ones that you approve when you use the product—that amount to: *we really respect your privacy and we're going to do whatever we want with your data.* If you're on OkCupid or Match.com, buy something from Amazon or Zappos, or activate the

OnStar system in your new car, you have handed your data over to the company.

What do they do with our data? They use it to develop detailed profiles of us—your "permanent record" in cyberspace—so they can market their products to you more successfully and deepen your engagement with them as customers. They may sell or exchange your personal data with partners and third parties. They may hand over your data in a merger or acquisition. They may feed your data to voracious AI large language models (LLMs) like ChatGPT that need increasing amounts of data to learn. For the most part, we don't know the specifics of any of this. We can't correct errors or provide context or give our approval or decline, other than checking the data policy box to use the product in the first place. And there may be no easy or convenient alternatives to these apps and services. We are hamstrung, with gluttonous algorithms running the show.

Because our profiles feed companies' abilities to personalize their products, each of us experiences cyberspace differently. Google the same term with your friends and each of you may get different responses and/or different rankings. A competitor—DuckDuckGo—did this experiment with keywords "gun control," "immigration," and "vaccinations," demonstrating that results varied widely.[6] Google's and other companies' browsers customize the results to be relevant to the perceived interests of users and advantageous to companies and advertisers.

None of this is surprising. Companies have used customized marketing for decades to find and engage customers. And it works. Companies continue to customize their marketing because it's good for the bottom line and, more often than not, convenient for customers. But cyberspace has taken customization and personalization to a new level. The detailed personal profiles companies amass include not just information we've provided them about our interests and preferences but information from a vast array of public sources, partner sources, and commercial data brokers. Information that can include your prescription medications, penchant for payday loans, sexual preference, political affiliation, and your and your children's locations.[7] User targeting is based both on direct information and information inferred from your vast data profile. If customized marketing was an art before, digital technologies have made it a science.

The specifics of your data profile can mean that you are offered higher prices online depending on a company's assessment of your ability and willingness to pay.[8] Customized pricing also isn't new—we are used to haggling for prices at flea markets and differential pricing for airline tickets, cars, and other items. But today's internet apps and services can exploit vastly more detailed profiles of a customer's ability to pay that include purchasing history, preferences, and financial resources—the Federal Trade Commission calls this "surveillance pricing."[9] We have the technology for virtually everyone to get a personalized invoice—an invoice that has used a mountain of data to calculate the highest price a company can get from each one of us.

All of this amounts to a cyberspace that is finely tuned to prioritize the people and organizations that make money from it rather than the people who use it. We expect this in the commercial marketplace. But we don't expect this from the essential goods and services we use as critical infrastructure. Roads, utilities, food, and other essential goods and services are expected to be safe, secure, affordable, and fairly allocated. We expect people to be prioritized as much as profitability.

Today's essential digital systems allow us to search, connect, communicate, purchase, process, and make decisions at an unprecedented level of scope and scale. But as they serve more and more as modern critical infrastructure, there is an increasing discrepancy between their success in the market and their failure to protect public interests, including our privacy, safety, and personal autonomy. The discrepancy—effectively a *risk gap*—occurs because, for the most part, we are building and managing technology to promote business interests, whereas critical infrastructure must be built and managed for public use. This risk gap is getting wider. Experience shows that letting the market alone drive the trajectory of the essential technologies of cyberspace has brought risks that many believe are already excessive.

IS CYBERSPACE TOO RISKY?

The average teen spends roughly five hours on social media each day.[10] In 2022, one-third of American teens were on social media "almost constantly."[11] What do they find there? Often a trip down an increasingly unhealthy rabbit hole. Gen Z writer Freya India describes the "conveyer belt" of social media from beauty tips to body dysmorphia (a mental disorder characterized by extreme preoccupation with a perceived flaw in one's appearance)—all to create the "Instagram face":

> Everyone on your feed starts to look perfect. You start editing yourself—smoothing your skin, reshaping your nose, restructuring your jaw. By the time you're 16, your Instagram face is very different from your natural face, which you've come to despise. And then the algorithms are introduced: your feed is no longer chronological but *customized.* Instagram now serves you not just photos of the friends you follow but of *influencers*—beautiful women from all over the world, selecting the ones that make you feel the most insecure. *You, with fuller lips! You, with a microscopic waist!!* Soon you get ads to fix your flaws: *Botox; fillers; Brazilian Butt Lifts!* . . . In this way, for many girls, this rewiring of their self-image, this pressure to alter their appearance, happened without them realizing it. It was gradual. Subtle. Drip-fed.

India continues,

> Algorithms act like conveyor belts. Show even the slightest interest, fear, or insecurity about anything—*hover over it for half a second*—and you will be drawn in deeper. Little by little, the algorithm learns what keeps you watching. And since the most negative and extreme posts get the most engagement, very often your feed will become an endless stream of content that makes you feel worse about yourself. You'll find yourself on a continuous conveyor belt of apps, products, services, pills, and procedures to fix you.[12]

What else do teens find on social media? Predators. As I write this, there is an increasing epidemic of sextortion of teenage boys—the online practice of soliciting nude photos and/or sexual favors and extorting victims for money by threatening to reveal them.[13] One of the victims was Jordan DeMay, a teen in Marquette, Wisconsin, who was solicited on social media in 2022 by scammers posing as a cute teenage girl. Things turned flirtatious; "she" sent him a nude of herself and asked for one of him in return. He sent one. The scammers then threatened to ruin his life by sending his picture to classmates, friends, and family. They asked for money and when he paid, they asked for more. Jordan texted that he wanted to kill himself. They told him, "Sure. I will watch you die a miserable death." Each text upped the ante. DeMay saw no way out. Within 24 hours of sending the nude, he committed suicide.[14]

In 2023, the National Center for Missing & Exploited Children received over 26,718 reports of financial sextortion, over twice the number as the year before.[15] That same year, the attorney generals of 33 states sued Meta, operator of Facebook and Instagram, for knowingly seeking to hook children on its platforms. The suits claimed that Meta had "designed psychologically manipulative product features to induce young users' compulsive and extended use," exposing teenagers and children to harassment, bullying, body-shaming, and sexual solicitation.[16] In a memo from 2016, Mark Zuckerberg, CEO of Meta, had directed executives to focus on getting teenagers to spend more time on Facebook. As per an internal correspondence, an "overall company goal is total teen time spent."[17]

In 2024, as a result of New Mexico's Operation MetaPhile, three men were arrested for targeting children for sex after they solicited state investigators posing as children on Instagram and Facebook.[18] The suit claimed that the targeting algorithms of Meta enabled the predators to identify children they would not have found on their own.[19]

It's hard to write about all of this dispassionately. It's every parent's nightmare. I have kids myself and as close as we are, I have no idea what they've experienced online—as teenagers and beyond. Predators are out there targeting children, teenagers, seniors, all of us. And cyberspace makes it easy. Companies use everything known about addiction to entice us to stay on our devices—to check our smartphones, to pursue likes, to share, to engage, to spend more time on their sites.

Facebook, Instagram, TikTok, X (formerly Twitter), and other companies have engineered social media to keep us interested, scrolling, and clicking. Apps and services are designed to drive surges of dopamine to the brain, to keep customers coming back over and over. Shares, likes, and comments trigger the brain's reward center, resulting in a high that's similar to the one people feel using drugs or gambling.[20] Notifications and alerts are designed to grab our attention and redirect us to our devices. Infinite scrolls—endless scrolls with no clicking needed—are designed to keep us online longer than we had intended. Experts call these strategies "behavioral cocaine."[21]

Moreover, all the time we spend online is currency—time that can be sold to advertisers, time that feeds the extraordinarily successful business model of offering "free" services on the internet to users in exchange for their data, their privacy, their attention, and often their mental health.

Studies have shown that prolonged and constant social media use can rewire the brain to continually seek out immediate gratification, leading to obsessive, compulsive, and addictive behaviors. This is particularly scary for children and teens, as adolescence is the second largest period of brain growth (the first is during the first five years of life).[22] Anecdotally, I have seen social media addiction in my own college classes. Unless I tell my students to "put away your devices," many cannot keep off them. They are distracted in class—scrolling, clicking, texting, dividing their attention. They have a hard time answering questions or engaging with the material. When I've told my students that we are going to have a device-free class, some of them literally cannot do it. They sneak their phones under the desk or behind their backpacks. They are addicts; they cannot function without their digital fix.

Is cyberspace too risky? Many experts believe that it is for children and teens. The risks are also increasing for the rest of us: 40 percent of internet users in 2021 experienced online harassment, and the prevalence of physical threats and sexual harassment was on the rise.[23] Anxiety and loneliness have been associated with excessive internet use.[24] Social media has evolved to be an insufficiently managed platform for election manipulation, the spread of misinformation, predation, and other harms.

Some think that the risks of cyberspace have already begun to outpace its many rewards. But the internet and its apps and services are profoundly embedded in the way we live; they provide the digital air we breathe. Although there have been calls to pause AI research, attempts to outlaw TikTok, and other reduction and cold-turkey solutions to reduce the risks of cyberspace, we can't stop innovation and we can't realistically get rid of technological risk by just pressing the OFF button.[25]

But decisions about how apps and services are designed, developed, used, managed, and abused are made by humans—us. And we can make different decisions, decisions that lower the risks of apps and services and elevate their protections—both when we create apps and services and when we regulate and manage them for the humans who use them. That's what we do with nondigital essential services and infrastructure. Once goods and services become essential—critical infrastructure—we begin to design, develop, and regulate them differently. We focus on making them safer, more secure, more reliable, more fair, *less risky*. We've done that with transportation, food, medicine, the electricity grid, and other essential goods and services. We've created societal controls that focus on public protections to reduce risks and improve benefits.

Our approach to nondigital critical infrastructure provides a playbook of strategies to reduce the risks of cyberspace. And that's where we can start to rein in the most egregious harms of our most essential technologies. But to effectively apply these strategies, we need a clearer picture of the protections critical infrastructure provides—and how the essential technologies of cyberspace fail us. We take a closer look at critical infrastructure in chapter 2.

2

BEASTS ON THE LOOSE: ESSENTIAL TECHNOLOGIES AND RISING RISKS

In early 2020, I was living in Boston for an eagerly anticipated one-year fellowship at the Radcliffe Institute, a fellowship that started me on the path to this book. When the COVID-19 pandemic hit and the fellowship abruptly went remote, everything changed. For me and many others, it was a deer-in-the-headlights moment, a time when it was hard to know how to respond and what to expect. In a single week in late March, I cleared out my office at Radcliffe, packed up my tiny apartment, and moved back home, worried and apprehensive about what was to come.

Over the next few months, the whole world would change and many lives would be disrupted, some permanently. Universities would go remote, and restaurants, offices, and stores would shut down. My son and his now-wife called off their wedding. In early 2020, the pandemic seemed to derail everything and there were no vaccines and insufficient science to explain it. Many worried about touching something that others had touched or breathing the same air and getting sick and short of breath. Millions would die of the disease.

During the pandemic, much of the world moved from on-site to online and cyberspace transformed the way we engaged with each other and our broader communities. I taught classes remotely and saw my friends on Zoom. Court hearings and religious services went online. In 2020, many students didn't throw their mortarboards into the air during graduation, get dressed up for their high school proms, or move into a dorm for their first year of college. It wasn't an option.

In all of this, cyberspace provided essential services so we could function—work, get information, learn, and access spaces online that were no longer available or safe in person. Our use of digital technologies brought both great rewards and increased risk. As many sheltered in place, cyberspace connected us and supported some semblance of commerce, school, community, and entertainment. The internet and its apps and services held up under our increased engagement, sustaining an uptick in traffic equivalent to Superbowl-levels every day.[1] At the same time, cyberspace's digital apps and services continued to hoover up our data in unprecedented amounts, gorging on our likes, clicks, and connections to feed online profiles with new information. Companies continued to prioritize market share over protections.

For example, ease of use upstaged security protections for Zoom meetings, making it easy for trolls to "Zoom-bomb" (intrude on) online gatherings to inflict lewd and racist content. Sunday school classes, Alcoholics Anonymous meetings, university courses, and a city meeting in Kalamazoo, Michigan, were disrupted with pornographic, racist, and inappropriate images. Public concerns about privacy and security prompted Zoom to tighten security and provide more user control, demonstrating that sometimes the market self-corrects without intervention.[2] But in many other cases—social media, AI-driven systems, surveillance—market forces alone have not been enough to promote the protections needed for widespread public use; proactive regulation or public intervention is warranted.

The pandemic clearly demonstrated that much of cyberspace, born of the largely unregulated wild west of commerce, has become digital critical infrastructure for today's world. For the most part, we take a "hands off" approach to its regulation and oversight, celebrating technology's innovation and success in the marketplace rather than ensuring that its public risks are minimized. Many economists say that we shouldn't overregulate the market, that regulation can limit innovation and skew capitalism. But what about the riskiest and most essential parts of cyberspace—digital critical infrastructure essential to societal functioning and well-being? Shouldn't we now take a more proactive, hands-on approach? Before we explore how in part II, it's important to specify more concretely just what we *do* expect from critical infrastructure.

WHAT WE EXPECT FROM CRITICAL INFRASTRUCTURE

Today's infrastructure is both virtual and physical. It's made up of "hard" infrastructure and "soft" infrastructure. *Hard infrastructure* consists of physical facilities—roads, dams, fiber optic cables, internet routers. *Soft infrastructure* consists of services and systems used to maintain the economic, health, cultural, and social workings of society, such as financial and educational systems.

Infrastructure deemed essential to the functioning of society—*critical infrastructure*—includes the electrical grid, the health care system, the internet, and many other infrastructure systems—both hard and soft. The Department of Homeland Security's Cybersecurity and Infrastructure Security Agency (CISA) recognizes 16 distinct sectors "whose assets, systems and networks, whether physical or virtual, are considered so vital to the United States that their incapacitation or destruction would have a debilitating effect on security, national economic security, national public health or safety or any combination thereof."[3] These 16 sectors are:

- Chemical Sector
- Commercial Facilities Sector
- Communications Sector
- Critical Manufacturing Sector
- Dams Sector

- Defense Industrial Base Sector
- Emergency Services Sector
- Energy Sector
- Financial Services Sector
- Food and Agriculture Sector
- Government Services and Facilities Sector
- Health Care and Public Health Sector
- Information Technology Sector
- Nuclear Reactors, Materials, and Waste Sector
- Transportation Sector
- Water and Wastewater Sector

CISA sectors include hard and soft infrastructure—both tangible facilities and the systems that drive them. The internet is claimed by two of these sectors: communications and information technology. Communications sector infrastructure includes broadcasting, cable, satellite, wireless, and wireline networks, as well as the critical voice, video, cloud, and data services that utilize them. The information technology sector is functions-based and includes virtual systems and networks that enable key capabilities and services, such as the internet's Domain Name System (the address book of the internet), often in collaboration with the communications sector.[4] Cyberspace doesn't operate without these facilities and systems. We saw that in 2024 when a routine security upgrade containing a software glitch from the company CrowdStrike caused an internet meltdown, crashing computers, bringing down hospitals and businesses, and grounding planes and trains around the world.[5]

Because the risk of failure of critical infrastructure from any of CISA's sectors could be catastrophic, we expect them to be built, regulated, and managed to reduce risk and incorporate public protections. Although each sector handles risk differently, some general characteristics and priorities for risk mitigation can help us understand why essential services and critical infrastructure need proactive management to support a functioning society and avoid debilitating harm.

These general characteristics and priorities are not surprising: Critical infrastructure is often *expensive to build* and *expected to last a long time*—we don't expect, for example, to replace nuclear reactors, dams, or undersea communication cables in the short-to-medium term. Critical infrastructure is often *hard to change*—the highway system, the electrical grid, or our health care system would be exceedingly difficult to restructure in fundamental ways. People expect critical infrastructure to be *created and maintained to be safe, secure, and resilient.* We fix potholes in roads, design power grids to minimize the impact of extreme weather and outages, and test drugs before they are released to the public. When things fail, *we expect to remediate problems*, even if it is costly to do so. Sometimes it is easier to fund the construction of new infrastructure than to sustain funding to maintain existing infrastructure—an economic challenge for both hard and soft infrastructure.

Critical infrastructure *may be risky in either its absence or its presence*—it's almost impossible for society to function when the internet melts down, or when critical facilities are malfunctioning, as may happen with aging dams and nuclear reactors that can put whole communities at risk. Importantly, people expect critical infrastructure to *promote the common good*—to be accessible to all people, to be fairly allocated, affordable, and nondiscriminatory, to be trustworthy, and to protect our health and well-being. Because of these common characteristics, *critical infrastructure is ideally developed, managed, and regulated to minimize its most egregious risks and with the public interest in mind.*

The public sector creates proactive interventions to ensure that critical infrastructure functions effectively—policy, law, regulation and enforcement mechanisms, standards and good practice, public-focused design choices, and other strategies we'll discuss in part II. The goal is to close the risk gap between infrastructure developed to promote business interests, and its widespread use by a vulnerable public.

Each of CISA's 16 sectors has its own approach to narrowing the risk gap: In the food and agricultural sector, we expect the food we buy in our grocery stores to be free from contamination. The US Department of Agriculture's Food Safety and Inspection Service (FSIS) was created to promote public health by preventing illness from meat, poultry, and egg products. In the first six months of 2024, FSIS inspections resulted in the recall of ready-to-eat charcuterie with possible salmonella contamination, beef products with potential *E. coli* contamination, and frozen ready-to-eat chicken pilaf with potential contamination from rocks—all to ensure that we don't get sick from what we eat.[6] The financial services sector has different interventions. When my wallet was stolen in New York City, the Fair Credit Billing Act of 1974, enforced by the Federal Trade Commission, was the reason I wasn't on the hook for $3,000 worth of charges I didn't make. This act protects credit card customers from paying charges they did not authorize, and the $50 maximum consumer liability is often waived, as it was in my case.

In whatever way critical infrastructure sectors narrow the risk gap, there is, of course, no guarantee that regulations will be effective and enforced, or that policies and protections will cover all risks adequately. However, in government, as in life, a little effort goes a long way. I expect my bank to be solvent, my bridge not to fall down, and my internet to keep working. For the most part, all of that happens. When problems occur—when the water in Flint, Michigan, was full of lead, when the Silicon Valley Bank went under in 2023, when the internet is disrupted and services go down—people expect there to be agencies, policies, standards, and laws and regulations in place to address the situation. For high-risk critical infrastructure, there often are. In modern civil society, we need those interventions to thrive: protections, remediations, and fixes that the market cannot provide on its own.

But what about cyberspace? Are we doing enough to narrow the risk gap for the essential services and systems on which we depend? Services and systems that may put people at risk when they are not secure or safe enough or fail to protect people's privacy. Services or systems in widespread use that may limit our choices or undermine people's mental or physical health and well-being. These systems and services are beasts on the loose, beasts that are beginning to proliferate in cyberspace, beasts that we need to contain. Many of these beasts are invasive—concentrating critical resources and consuming the competition, especially in the rapidly growing and resource-intensive domain of AI.

We will need to create interventions to control the risks of these technological beasts. And we will need to design, develop, deploy, and manage them differently than we do now. Cyberspace will need to look a little less like Jurassic Park and a little more like the San Diego Wild Animal Park.

The good news is that in cyberspace both risks and remediations are human-made—we can choose how to control our beasts on the loose. We turn to proactive strategies for controlling the risks of tech in part II. Before we do, we need to consider which technologies it's important to control.

WHICH CYBERSPACE TECHNOLOGIES ARE ESSENTIAL?

We start with the digital systems that CISA designates as essential. For all the information technologies that drive its critical infrastructure sectors, CISA is particularly focused on cybersecurity, and reducing cyber risk is a continuing battle for all sectors. Consider health care.

Digital technologies are transforming medical science and care. Electronic medical records have become digital health logbooks, living in the cloud, ideally read and accessed from anywhere. Artificial intelligence is being used to help read X-ray images to detect potential tumors, expanding our capacity to detect minute anomalies. Connected implanted medical devices—pacemakers, insulin pumps, and the like—are collecting critical data for patients and doctors to manage serious health conditions.

CISA recognizes that digital technologies used in health care have vulnerabilities that can create new risks and require proactive intervention, sometimes in the form of recalls. In 2017, the US Food and Drug

Administration issued a voluntary recall of almost half a million of St. Jude Medical's connected pacemakers "to reduce the risk of patient harm."[7]

The St. Jude Medical pacemaker system, like many other connected medical devices, is essentially a minicomputer that communicates over the internet to physicians and hospitals. It employs a device implanted under the skin that collects and transfers cardiac data to home transmitters. The transmitters in turn send the information to doctors and hospitals via phone, cellular, or broadband networks, completing an integrated system that can alert patients and doctors to heartbeats that are too slow or too fast.[8] Such systems have all of the advantages of thorough and expeditious feedback between patient and doctor. They are also more vulnerable to compromise—their digital components provide a bigger "attack surface" for bad actors and cybercriminals.[9]

In a report in 2015, MedSec, a cybersecurity company specializing in medical devices, alerted the public to a security vulnerability of the subsequently recalled St. Jude Medical pacemaker: the device could be hacked. Security vulnerabilities provided a digital trespasser the ability to modify the program to deplete the device's battery, change its commands, or administer inappropriate pacing or shocks—potentially killing the patient.[10]

No one died as a consequence of these security vulnerabilities and the system was recalled, but this kind of threat was, and is, real. The idea of hacking a connected medical device and killing its human host is regularly the stuff of entertainment—fiction that describes a realistic danger so frightening that Vice President Dick Cheney had the wireless feature on his pacemaker disabled as a matter of national security. "I was aware of the danger [and] I found it credible," said Cheney when interviewed about his decision on the TV program *60 Minutes* in 2013.[11]

But connected technologies are an increasingly essential part of today's delivery of health care, diagnosis, and treatment. And we will always be behind the 8-ball when it comes to their cybersecurity—it is virtually impossible to build completely safe complex systems and it will always be possible to break them—all complex systems have weak points, vulnerabilities that can exacerbate risks when something goes wrong. CISA's investigations, analyses, and advisories for medical devices and systems are critical to continue to ensure safer and more empowering technology as security vulnerabilities are exposed and new technologies are developed.[12]

The expectation that we must minimize and mitigate dangerous risks to protect the public is baked into our approach to critical infrastructure. It is not an expectation we have about our most indispensable digital technologies. Although tremendously imperfect, the existence of CISA, its recommendations and advisories, and the attitude that *the public sector has some responsibility to manage the technologies that drive critical infrastructure to reduce human risk* is fundamental to our well-being. We must now begin to apply this mindset to cyberspace.

Beyond the information technologies that drive CISA's critical infrastructure sectors, many other systems and services of cyberspace have become indispensable to modern life. For example, it is impossible to imagine navigating the modern world without *online information*, so much so that "googling" has now become synonymous with the search for information. Search engines like Google, Bing, Baidu, and DuckDuckGo deliver—and shape—the vast majority of the information we consume.

We have become reliant on *e-commerce and online platforms*—Amazon, Expedia, Zillow, and many others—to drive purchases of everything from pet supplies to plane tickets to housing. These services determine what we see and can influence access and pricing. They are convenient commercial online enterprises, often widely preferred to on-site enterprises, and sometimes the only way to function when on-site alternatives are absent, as we saw during the pandemic.

Many services and platforms in cyberspace are essential for *connection and engagement*. Over the last decade, social media—Facebook, TikTok, LinkedIn, Twitter/X, Reddit, Truth Social, and others—has become the place where major announcements are made (resignations; hirings and firings; corporate, government, and personal news, etc.) and where active communities are built and maintained. Social media was critical to communication during the Arab Spring, the January 6 insurrection, and the

COVID-19 pandemic. It increasingly serves as a primary source for critical information (and misinformation), much of which is inadequately curated and controlled if it is to be considered trustworthy.

Online collaboration and communication tools have also become essential technologies. The pandemic changed the organizational model for many businesses to focus on remote work, and many of those businesses have become hybrid—accommodating both remote and on-site workers. Cyberspace services like Zoom, Teams, Microsoft 365, Google Workspace, and others have made this easy and allow anyone with a connected device and a decent internet connection to work, collaborate, and share project infrastructure remotely. Communication and messaging tools like WhatsApp, Signal, Telegram, Facebook Messenger, Slack, and others keep us chatting, connecting, and communicating in cyberspace.

Essential services and systems also provide *foundational infrastructure* for a robust, thriving society, including systems supporting education, elections, emergency services, and other domains. As we saw with CISA sectors, there is little in our complex, modern society that does not increasingly incorporate—and often rely on—the tools of today's information technologies.

Although cyberspace is full of essential technologies, it is important to note that *the level of risk of essential digital systems and services is not always high*. Risk depends on the way users engage with technologies, their potential vulnerabilities and outcomes, and the policies of their providers.

Determining this isn't always easy. Both the use and the level of risk of digital technologies can vary. For example, social media was neither essential nor high risk at its outset but has increasingly become both—a beast on the loose. This became especially clear when social media served as a primary information resource for health information during the pandemic, underscoring the need for more public scrutiny to determine what is accurate and what is not.[13]

During the pandemic, social media was saturated with advice about COVID, advice that was inconsistent and sometimes potentially dangerous: Some posts claimed that vaccinations were being used by the government to microchip the public. Some posts claimed that vaccines offered protection against severe disease and lessened the death toll of the virus.[14] Some posts claimed that injecting bleach or isopropyl alcohol into your body could be used to combat the disease.[15] What were users to believe? Without vetting health information for accuracy (which most users did not do[16]), the (mis)information found on social media could reinforce any one of a variety of narratives about the pandemic—narratives that might increase or decrease users' risk of becoming severely ill from the disease or a proposed treatment, narratives that could have a real impact on the health of the people users came in contact with.

Identifying which technologies are essential—digital critical infrastructure—provides a foothold for reducing the public risks of cyberspace. To make tech better, we can then work to lower the risks of the essential technologies with the greatest potential for harm first. This brings us to our next challenge.

WHICH ESSENTIAL TECHNOLOGIES ARE HIGH RISK?

Technologies put us at risk when we are not able to engage in society freely, fairly, and without undue manipulation. They put us at risk when they make it easy for others to exploit or prey upon us, or when they mislead us or make us unsafe. Many of the technologies you and I engage with every day are breeding grounds for these kinds of risks.

High-risk essential technologies include those that underlie the *search and delivery of information*—your search engine and browser, for example. Your search engine and browser don't just deliver information. They choose what to show you based on their own business interests, AI recommendations, and personal information collected about you. Googling is not the same as searching for information in a library, where you control the way you go about your search, and the library provides an organizational structure for content. Online you have limited control over what your browser and search engine serve up and little transparency and understanding of the algorithmic priorities that guide those decisions. To increase benefits and reduce the risk of user manipulation and discrimination, browsers and search engines should provide greater transparency, user control, and privacy protections as they deliver the information we consume.

Technologies for *connection and community*—for example, social media—have become increasingly essential for modern work and increasingly prevalent for community engagement. But the risks of participation have increased, putting many social media platforms in the high-risk category. Users often have limited control over which posts they see and how information is disseminated, vetted, and shared. Cyberbullying, manipulation, and addiction all exacerbate risk. Proactive intervention is needed to promote healthier community engagement and help users assess the trustworthiness of information. For some platforms and in some jurisdictions, social media content and contributors may be censored based on political perspective, national identity, human rights advocacy, and other considerations. Account suspensions, takedowns, and other forms of algorithmic suppression may be used to limit the fair and free expression of information rather than to weed out sources of misinformation. Which platform you use, what country you live in, and what is considered acceptable or not all impact how social media mediates the information that is often critical to full participation within communities.

Lower-risk essential technologies include services that underlie our *ability to communicate and collaborate online*, from email to messaging apps to video conferencing systems. Broad use of these systems is fundamental to full participation in society—imagine today's world without email and messaging. Invasion of privacy, phishing attacks, identity theft, and intrusion in these systems put users at risk, but broad incorporation of password protection, encryption, two-factor authentication, and other techniques and interventions have lowered these risks for many users. Even so, it is often easy to scam users and exploit systems. It is imperative that companies and users alike stay vigilant.

Better tech will need to start with the tech that matters most and has the greatest potential to harm us, meaning the tech with the highest risks. It would be crazy, counterproductive, overkill, and unrealistic to try to create interventions for every technology in cyberspace. But it is equally crazy—and increasingly dangerous—to let the market take the lead while the public sector creates protective interventions for too few of them.

And then there's AI. Many of today's and tomorrow's technologies will be driven by artificial intelligence, bringing increasing capability and a new source of potential risk to cyberspace. Although AI has been an important focus in the research community for decades, its increasing capabilities and commercial value were largely ignored outside of the tech community until the public release of ChatGPT in late 2022.

ChatGPT's release served as a wake-up call to the broader public as to the potential, promise, and perils of AI as a disruptive force. Articles abounded in the media: ChatGPT was capable of passing an MBA exam at the Wharton Business School.[17] After a prolonged "conversation," Bing's chatbot tried to convince a journalist to leave his wife.[18] Students could use AI to do their homework for school. A lawyer used ChatGPT to create a brief that cited nonexistent legal cases.[19] Improvements to the original large language models (LLMs) used for ChatGPT and other AI systems—such as Claude, Gemini, DeepSeek, and others—are increasingly making them more accurate and more powerful. They are also exposing both people and the planet to new risks as AI begins to colonize cyberspace.

When we think about the technologies we consider essential, a growing number will be driven by AI—cars, consumer services, recommendation and decision-making systems, and many others. Proactive interventions from the public sector will be needed to ensure that AI-driven systems and services do not put us at extreme risk or disadvantage. Interventions such as the European Union's AI Act, set to go into effect in 2026–2027, provide an example of how the public sector can proactively manage the risks of AI.[20] (We describe the AI Act in more detail in chapter 4). This groundbreaking legislation is the first comprehensive legal framework that aims to couple public risk with public scrutiny—the more dangerous a system could be, the more public oversight is required to protect the people who use it. In effect, the AI Act provides a mechanism for managing high-risk AI as digital critical infrastructure.

WHO IS MINDING THE RISK GAP?

Today, there are insufficient public protections for digital critical infrastructure, leaving the essential technologies of cyberspace largely unregulated and unpoliced, especially in the United States. For many of the highest risk essential systems and services, the expectation that the public sector has some responsibility to manage them to protect the public does not yet apply.

There are many reasons for this: Some companies don't do a comprehensive risk analysis or minimize the importance of one in the face of business interests. In their rush to lead in the market, they seek to "move fast and break things," as Facebook did early on, streamlining both the costs and time for product or system development and skimping on public protections. In addition, companies are often unaware of the dangers that come with broad use. It may take a period of adjustment—William Ogburn's cultural lag from chapter 1—to recognize the most severe risks and rein them in.

The risk gap is also affected by the social environment in which products and systems are developed. In the United States, we are often loath to limit companies because we do not want to curb innovation in a capitalist system—an argument that is regularly used by the powerful and well-resourced to prioritize the rights of corporations to pursue profit over the well-being of their customers. Some essential parts of cyberspace have become a political football: For example, net neutrality—the policy that guides whether the internet is considered a utility and who can profit from its provision—is not law and is implemented or rescinded based on the politics and priorities of each administration. And finally, the public interest in cyberspace is often deprioritized in our efforts to come in first in global trade, negotiate between warring political parties, appease industry, and deal with the latest urgent or pressing priority. Some of the most essential systems and services in cyberspace remain an ungoverned wild west.

We need a full court press to narrow the risk gap for digital critical infrastructure. For humanity to thrive, especially in an era where AI will increasingly call the shots, we must proactively ensure that technology empowers rather than endangers us. Without proactive interventions, the technologies with the biggest risks in cyberspace—the beasts on the loose—can wreak havoc in our lives, despite their benefits and potential to promote societal advancement.

Two examples demonstrate that proactive intervention can go a long way toward promoting the public interest. Both focus on inequity and essential technologies. The first focuses on the risks of essential technology that is not accessible to all; the second focuses on accessible technology that is not equitable for all. Both are examples of areas where proactive interventions are needed to help narrow the risk gap.

Closing the Risk Gap in the Digital Divide

During the COVID-19 epidemic, schools and universities were some of the first organizations to migrate from on-site to online. In 2020, I was able to teach my students remotely but many struggled—sharing a computer with working parents and/or siblings, in homes with no private place to participate in class, or at the mercy of inconsistent internet connectivity. These students were the lucky ones. The pandemic exposed and exacerbated a digital divide that prevented many students from attending school remotely on a regular basis or at all: Students who were disproportionately from poor, rural, or marginalized populations, students who had unstable or unsafe housing situations, students who were homeless.

In 2019–2020, over 124,000 elementary, middle, and high schools shut down, moving 55 million students to remote learning with only 24 percent returning to full-time in-person classes the next year.[21] The disruption in school was traumatic and broadened already concerning levels in social development and learning among the students most in need of help. "The necessity of remote learning in 2020 [turned] the homework gap into a chasm. . . . a cavernous divide that will harm students without robust reliable broadband at home for the rest of their lives," observed the think tank New America.[22]

In 2020, 50,000 of the 55 million students who moved to remote learning lived in Oakland, California. Compared to nearby San Francisco and Silicon Valley, Oakland is under-connected and under-resourced—at the beginning of the pandemic, only 1 in 4 of Oakland's 50,000 elementary, middle, and high school students had a device and decent internet connectivity. For low-income students, this was 1 in 8. Some students attended school sitting on a bench outside the library hotspot trying to write a five-page paper for high school on a cellphone. One mother had four children in the school system, one loaner device to share between the family, and consistently spotty internet. She was not alone.[23]

Two years later, 98 percent of the children in Oakland's Unified School District could attend school remotely. Oakland distributed over 36,000 computers, most of which students could keep, and created 11,500 hotspots to help the kids connect.[24] How did this happen? The digital divide is a collective problem. It is also a human-made problem, and educators and community members in Oakland came together to develop a collective, human-made solution. Public, private, and community organizations created a coordinated #OaklandUndivided initiative to address the problem from all angles.

The mayor's office and the Oakland public school district worked with community nonprofit Tech Exchange to develop the infrastructure to assess who had sufficient access to a computer and who had connectivity, and to buy the computers and hotspots needed for students to go to school remotely. Tech Exchange's public-facing TecHub provided free tech support and training on site at public housing complexes. Its staff worked with the Latino Education Network and the Oakland branch of the NAACP to target the Latino and Black populations that are prevalent in Oakland. Community members launched OAK WiFi to increase connectivity in the area and reached out to the state to expand local broadband infrastructure. This involved fundraising from both the community and big donors like Jack Dorsey, then-CEO of Twitter and Square; the development of infrastructure to track who had a computer and connectivity and who did not; and the mobilization of citizens and groups needed to distribute computers and hotspots and to help students connect.[25]

Sometimes it was hard to reach the families of students still left out. The district engaged community workers who spoke Spanish and Mam, a Mayan language, to reach out to families whose kids were not connected. Some families had trouble picking up the computers as parents were essential workers. Some had trouble because of housing instability and shifting addresses. Some had trouble because they were recent immigrants to the area and had not situated themselves and their children. It took special efforts by dedicated community workers to find every possible student and give them the tools they needed to go to school.

The narrowing of the digital divide for school children during the pandemic did not solve all of Oakland's digital divide problems. Shifting politics, priorities, and economic challenges require persistent effort to sustain the gains made with #OaklandUndivided and to bring broadband to the area, which can help provide a more permanent solution. Community groups and local government continue to work to address this challenge.

But public interest efforts like #OaklandUndivided can, and do, make a difference. They can remediate the risks and disadvantages that come from inadequate access to essential services and systems of cyberspace. The digital divide didn't just happen during the pandemic, and it isn't just about education. Then–Oakland mayor Libby Schaaf defined the problem concisely: "It was so much bigger than just education. It was literally survival for Oakland families. Technology was the lifeline to information."[26]

Many rural, tribal, and underserved communities remain on the wrong side of the digital divide. In 2023, 42 million Americans still lacked broadband access.[27] The increasing use of the internet as critical infrastructure and connective tissue on which essential systems and services rely makes universal broadband an urgent priority. Federal programs have been created to bridge the affordability and availability gap for many Americans, but support varies by administration and much still needs to be done; many Americans are still left out in the digital cold.[28]

Author William Gibson famously said, "The future is already here, it's just not very evenly distributed."[29] #OaklandUndivided's efforts show that if we *prioritize* and *actualize* investment in the public interest and aggregate efforts, power, and resources across multiple scales, we can more evenly distribute the infrastructure on which we depend. And we can dramatically narrow (and maybe ultimately close) the digital divide. This is vital to ensure that the internet serves as critical infrastructure for all.

Identifying the Risks of Essential Technologies: AI and Discrimination

Joy Buolamwini was a grad student at MIT working with an AI-powered facial detection system when she made a disturbing discovery.[30] Buolamwini is Black, and when she used her own face with the system, it came up (literally) with a blank. When Buolamwini put on a white mask, the system recognized her and could make out her facial contours. Her subsequent research demonstrated that common commercial AI systems, trained on commonly available data sets, often of predominantly "pale males"—light and white-skinned men—did a much poorer job of recognizing people of color and women, meaning that the percentage of errors in identification was much higher for those groups. Such systems are increasingly used in predictive policing, recruiting, authentication and access, and many other areas to drive decisions.

The failure of these systems to accurately identify people of color, and the use of these systems without sufficient human oversight has led to high-profile arrests of Black individuals including Porcha Woodruff (falsely arrested when she was eight months pregnant when her facial "match" was not visibly pregnant), Robert Williams (arrested in his driveway in front of his two young children), and Randal Reid (arrested for a crime committed in a state he'd never been to).[31] For these individuals and others, biases and inaccuracies in facial recognition systems coupled with inadequate human investigation can have disastrous social consequences.

In a way, this kind of algorithmic bias is not surprising. The training sets we use to help many facial recognition systems learn are often not representative of the population the systems will be asked to recognize.[32] Why would we expect a system trained on mostly middle-aged, adult White men to provide accurate results for people of color, women, and older Americans? AI ethicist Timnit Gebru and Buolamwini created a more representative and diverse training set of adults from Parliament members in countries all over the world, a world—our world—in which the majority population in many countries is people of color.[33] Using benchmarks such as theirs in algorithmic auditing systems as well as during the design, development, and testing of facial recognition products would help ensure that these systems better identify and serve the public in advance of their widespread deployment and use. There are currently few incentives to do this, but there could be more.

When the tech that provides essential services to law enforcement, businesses, and the government is discriminatory, it impacts how groups are treated, what opportunities they are offered, and what risks they face. Because of Buolamwini's research, some companies reengineered their systems or moved away from facial recognition altogether, but others did not.[34] In a subsequent facial recognition vendor study, the US National Institute of Standards and Technology (NIST) tested 189 algorithms from 99 developers—a majority of the industry. The systems were tested on images of 8.49 million people from databases provided by the State Department, the Department of Homeland Security, and the FBI. The results showed many kinds of inaccuracies, including more false positive matches for Black females (similar to Buolamwini's results) and Asian and Black faces compared to the number of false positives of images of White individuals.[35]

In practice, the accuracy of applications that use commercial facial recognition systems to identify people of varied gender, age, and racial backgrounds depends on multiple factors. These factors include the model at the heart of the facial recognition system, the data it's fed, whether there is a "human in the loop" monitoring the results, and how the facial recognition system is incorporated into an application. The systems tested by NIST and in Buolamwini's studies are commercial products—anyone can use them and companies that do may make application decisions based on technology compatibility, cost, existing partnerships, and other criteria as much as on their NIST scores.

NIST continues to do critical work in assessing the accuracy of facial recognition algorithms, developing potential standards, and providing guidance to industry and government agencies on facial recognition systems. We can build on this as a foundation in our pursuit of better tech. To equitably protect all communities within the public, we should expect high-risk applications using facial recognition to be audited for accuracy and to meet specific standards. We should expect that the use of facial recognition systems is carefully monitored by humans who can vet the accuracy of their results and reduce the incidences of false arrest, misidentification, and other harms. As we continue to learn more from studies like Buolamwini's and NIST's commercial algorithm assessments, we should use this knowledge to improve how facial recognition systems are developed and how they are used, so as to reduce their risks and increase their benefits.

NARROWING THE RISK GAP FOR ESSENTIAL TECHNOLOGIES

The creeping normality of risk in cyberspace is imperceptible and largely ignored. Data breaches, phishing, and sextortion are regular occurrences. Bots are everywhere. Trolling on social media is rampant. Many of us have personal information or social security numbers readily available on the dark web. Our trust that technology will be accurate, reliable, and secure is low and diminishing.

An old fable tells of putting a frog in a pot of water and raising the heat so gradually that the frog is not aware that it is being boiled. As the risks of cyberspace compound and as AI helps raise the temperature, will we know when the risks of cyberspace are too many? When the water will be too hot? Are we already starting to boil?

Boiling the frog is a fable, one that resonates with many of us who think the water is already too hot. Biologists tell us that the fable—while good at illustrating the creeping normality of risk—isn't accurate about frogs. Experiments show that frogs put in boiling water will die and frogs in water being gradually heated create their own proactive intervention—they jump out of the pot when the water is too hot.

It's not clear how we would realistically jump out of the cyberspace pot, but unlike the allegorical frogs, we have the capacity to turn down the heat. Both the risks and remediations of cyberspace are created by us. With much effort and persistence, we can begin to turn down the heat for the technologies we consider most risky and most essential—our digital critical infrastructure. We saw that when #OaklandUndivided worked to narrow the digital divide during the pandemic. We saw that when Joy Buolamwini's work alerted the public to bias in facial recognition and the National Institute of Standards and Technology tested and developed standards for commercial facial recognition systems. We see that in the AI Act passed in Europe to reduce the risks of smart systems and in laws passed in California to protect consumer privacy, both of which we will discuss in future chapters.

None of these are perfect solutions but they help lower the temperature in the cyberspace pot and reduce risk. Such efforts come from both the bottom up—individuals and organizations and communities—and the top down—governments and powerful multinational companies; everyone has a part to play. Part II helps us develop a public interest mindset by giving us strategies for big and small efforts to reduce the worst risks of our most essential technologies, strategies that can help us ensure that digital critical infrastructure is better created and managed to put people first, strategies that can help us turn down the heat.

II

MAKING AND MANAGING BETTER TECH

PROLOGUE, PART II

"It takes a village to raise a child" has its roots in an African proverb and means that many people play a role in creating and maintaining a safe and healthy environment to nurture a child.[1] The phrase is often used to illustrate the fact that it takes many types of efforts to nurture successful projects, products, organizations, and social issues, and to keep them on the right track. It's certainly true for cyberspace. At every stage of the technology life cycle—innovation and conceptualization, design and development, deployment and use, management and governance, and even disposal and end-of-life, many socially responsible efforts can increase the benefits and reduce the risks of the essential products, services, and systems of cyberspace.

How to keep social responsibility at the forefront? Consider cars, part of the Cybersecurity and Infrastructure Security Agency's transportation systems critical infrastructure sector. At every stage of the automotive life cycle, proactive intervention has been used to make cars safer (reducing the risk to humans) and more environmentally sustainable (reducing the risk to the planet). Many efforts and people over many years have proactively nudged the automotive industry to avoid, reduce, mitigate, remediate, and manage human and planetary risks.

Some history: Although gas-powered motorized vehicles were invented in the late 1800s in Europe, they were not in widespread use until the early 1900s in the United States.[2] By the mid-twentieth century, cars had become essential to the functioning of society. Cities were planned around driving, not walking. Urban sprawl, suburbs, gas stations, drive-thrus, and Google Earth are all ripple effects of the ubiquity of automobile transportation. Today the automobile industry is a major driver of the global economy. It employs millions of people and is fundamental to our way of life.

Automotive transportation is critical infrastructure. It is also high risk. Our increased mobility, prosperity, and lifestyle flexibility have also given rise to highway accidents that have killed millions of people and environmental pollution with the potential to choke the planet. Over the last 100 years, persistent and proactive efforts by the public sector, auto industry, research community, and citizens have sought to lower the human and planetary risks brought on by ubiquitous automobile travel with much success. How?

One innovation that has improved the environmental sustainability of automobiles has been to change the way they are powered—by replacing engines based on gas combustion with electric power. Electric vehicles (EVs) have paved the way to a more environmentally sustainable transportation future, but it's taken time. Although EV technology had been in development for decades, electric vehicles did not see widespread investment and adoption until the 2000s, when consumer demand, environmental concerns, and design innovations converged in a perfect storm. By 2023, nearly one in five cars sold was an electric car.[3] In the next decades, the majority of the cars on the road are expected to be electric vehicles.

Why is this better tech? In the early 1950s, CalTech researcher Arie Haagen-Smit, now known as the "father of pollution" (in a good way . . .), was studying the damage of smog on plants in the Los Angeles Basin and the impact of smog on human health. His work showed something we take for granted today but didn't know then—that the invisible exhaust from tailpipes on gas-driven cars plays a key role in the photochemical reaction that creates smog.[4] Smog so thick that the carbon dioxide from car emissions and industrial manufacturing has contributed to a pollution dome over the LA Basin that is easily detectable by satellites in space.[5] Haagen-Smit's careful research enabled people to understand the causes of smog and helped lay the groundwork for legislation such as the Clean Air Act (originally passed in 1963) and standards that contribute to the reduction of emissions that pollute our skies.

Public-focused research, innovation, and legislation helped nudge the automotive industry to be more sustainable and promote human health. What else helped? In 1965, Ralph Nader published the book *Unsafe at Any Speed*, alerting the public to the defects and dangers of the Chevrolet Corvair.[6] The book, and the recognition that vehicle accidents were the leading cause of death for Americans under 44, spurred the first seatbelt legislation and the establishment of the National Highway Traffic Safety Administration (NHTSA), a public agency whose mission is to promote transportation safety. Since the founding of the NHTSA, vehicle safety, security, and more recently, privacy have been critical priorities for the transportation sector. NHTSA's proactive approach to performance standards, education, and partnerships has helped make safety a top priority in the auto industry and saved many, many lives.

No one can, or should, rightly claim that automobile transportation is completely safe and good for the environment or that the efforts made thus far are enough. But they've helped a lot. Vehicle transportation is critical infrastructure, and over the last 100 years, researchers like Haagen-Smit, advocates like Nader, companies like Nissan, Tesla, and GM (which helped commercialize and popularize modern electric vehicles), government agencies like the NHTSA, industry standards bodies, and many other efforts and proactive interventions have promoted automobile safety and environmental sustainability. Over time, complex systems *can* be nudged to better benefit and reduce risk for the people who depend on them.

In part II, we focus on major strategies that can be used to reduce risk and put people first in cyberspace—responsible tech design, development, and deployment (chapter 3); public interest–focused law, policy, and regulation (chapter 4); enforcement and effectiveness strategies (chapter 5); and the support of critical people in the loop—advocates and educators (chapter 6).

The last 100 years of transportation interventions and remediations, better designs, and more focus on safety have shown us that things can get better. The next decades will require us to continue to promote the public interest and planetary sustainability as AI disrupts the world of transportation. Experts assert that in the last half of the twenty-first century, self-driving cars will dominate our roads. They will change our transportation economy, built infrastructure, and energy usage, and they will impact our safety, personal privacy, and lifestyles. And transportation is just the tip of the iceberg for the changes AI is bringing to our world. How do we promote the public interest in the context of rapid and disruptive technology evolution? We consider how to prioritize the public interest in a cyberspace increasingly dominated by AI in part III.

There are many ways to reduce risk. The strategies described in the next chapters are applications of a public interest tech mindset and provide a playbook of sorts to advance the public interest for the essential and high-risk technologies of cyberspace. Just as our persistent and proactive work over many years has made transportation safer, these strategies can work synergistically and over time to help us build the beneficial cyberspace we deserve.

3

AN OUNCE OF PREVENTION: RESPONSIBLE DESIGN, DEVELOPMENT, AND DEPLOYMENT OF TECHNOLOGY

Critical infrastructure is designed and developed to *work*: to function without breaking, to minimize risk for the people who use it, to be sustainable over the long term. The way we create and use our most essential cyberspace technologies should also be driven by these goals.

The way we design and develop things sets the stage for their benefits and risks. Architects design stairs with standard treads and rises so that you don't trip and fall. Highway engineers plan roads with materials that give your car traction so you are less likely to skid. Software developers incorporate password protections to add an additional level of security and privacy to your user account. Putting public health, safety, privacy, and security first when designing and developing technology provides an ounce of prevention. When we reduce the potential for risk and malfunction, it's easier to avoid having to remediate problems with a pound of cure.

The creation of technology is the result of a myriad of human decisions. We build tech *responsibly* when we minimize its risks to humans and when we align design and development priorities and processes with the best interests of people and the planet.

Although no one sets out to create technology *irresponsibly*, a responsible approach to creating and using tech involves *intention* and *follow-through*—intention to actively prioritize and incorporate human and environmental protections; follow-through by teams and organizations even when public protections might conflict with maximum profitability, functionality, or speed to market.

There are many ways to promote tech responsibility. In the sections below, we explore responsible technology design, development, use, and operation. The fact is, *we know how to do this; we just have to make it important enough.* The more prevalent technology becomes in our lives, the more critical it is to create and utilize tech responsibly. Doing so is an investment in our present and our future.

RESPONSIBLE DESIGN

New technology starts with a problem and the conceptualization and design of its solution. Interestingly, you never start with a clean slate. The development of new ideas always involves working around constraints—the laws of gravity, the properties of materials and systems, laws and regulations that allow you to do some things and not others, your own sense of ethics and fairness, your desired outcomes. Among the choices one makes during conceptualization and design is whether and how to factor in the potential impact your choices will have on society: Will your approach work for everyone or best for a particular group? Will it safeguard the environment or exacerbate problems? What are the rewards and what are the risks of various design options?

The lens of social impact becomes especially important when we are dealing with technologies that are likely to be used by many at scale, and likely to become essential. Asking questions about societal impact and using what we find to guide beneficial, low-risk solutions to potential problems is a responsible design strategy that can promote human benefits and reduce human harm. As companies, the government, and citizens are finding, it's much easier to avoid serious risks than to mitigate them.

Why? It's often difficult to refocus or retrofit mature products, services, and devices to incorporate public protections. Once products are used by many and are part of the complex ecosystem of cyberspace, we're kind of stuck; for example, it's unrealistic to reverse engineer the internet to support more stringent cybersecurity or TikTok to be less addictive. And most risk mitigations won't have the impact of a more risk-averse technology redesign, something that is prohibitively expensive for today's most mature essential technologies. The bottom line: *Technology is more likely to reduce risks for its users if the public interest is part of its design from the outset and not an afterthought.*

Consider self-driving cars—a transformative convergence between the increasing capabilities and potential of AI and the future of automotive transportation. Tomorrow's driverless cars offer a working example of the challenges and opportunities of designing an essential technology that can reduce pollution and promote planetary sustainability, and highlight how a public interest tech mindset can guide innovation.

Designing for Sustainability: Will Self-Driving Cars Be Good for the Environment?

What happens when AI does the driving? Today's cars are on a trajectory to answer this question with every new generation of vehicles incorporating more and more AI features that perform the functions of a human driver. Today's "smartest" cars can self-park, have auto-pilot systems that drive themselves at highway speeds, incorporate autonomous collision avoidance, and more. Going forward, our cars will continue to get "smarter" and more independent. Experts predict that by 2045, half of the new cars for sale will be primarily AI-driven, and most will be electric.[1] Today, it is impossible to buy a new car without some AI capabilities. By the latter half of the twenty-first century, self-driving cars are expected to be everywhere.

Will the self-driving electric vehicles of the future be good for the environment? You might think this is a trick question. Gas-powered cars have created smog all over the planet, but the AI-driven vehicles that will be ubiquitous in 2050 and beyond are slated to run on electricity. How could self-driving cars be bad for the environment? Depending on how we design the AI systems that drive over the next decades, they might be. The systems that propel self-driving cars will have many impacts on the environment—on air quality, on the materials needed to manufacture them, and on the in-car and built infrastructure needed to support them (think electric charging stations, persistent internet access for the roads cars travel on, highway speeds controlled by algorithms). AI-driven vehicles can be designed and developed to be beneficial for the environment by planning ahead now, or we could just leave it up to the market. It's up to us.

How would self-driving electric vehicles pollute? Emissions come from the entire life cycle of the car—manufacture, operation, and end-of-life disposal as recyclables or e-waste. Electric vehicles mostly produce emissions during manufacturing and disposal and also when the electricity that powers their batteries is generated. Today, electricity is still largely generated by burning fossil fuels; we will need to explore low-emission design alternatives for creating and decommissioning electric cars.

But it's not just vehicle and electricity production. A portion of the energy consumed by self-driving cars will be used to power on-board computers and sensors needed for AI—the systems that replace the human functions of modeling the environment (sensing and planning) and operating the car (driving). Systems that make self-driving cars "data centers on wheels" and will still produce emissions.[2]

A few years ago, MIT PhD student Soumya Sudhakar and her colleagues were curious about whether and how much self-driving cars will pollute the environment when they are everywhere: How much carbon dioxide would a future fleet of self-driving cars emit? They modeled a 2050 scenario where one billion vehicles, approximately 95 percent of which were autonomous, operated for one hour a day. (The study made conservative assumptions about the 2050 scenario—that the computational workloads of self-driving cars would keep pace with their growing algorithmic complexity and that decarbonization—removal of carbon dioxide—would continue at a business-as-usual rate.)[3]

The study indicated that *without proactive planning and design alternatives, emissions from self-driving electric cars could spiral out of control.* In particular, the study showed that this global fleet of self-driving electric vehicles will need *a lot* of power—to process data from cameras and sensors, to model the environment, and to make decisions about vehicle operation. Generating that power could create greenhouse gas emissions comparable to that of all data centers in 2018.[4] The bottom line: *Self-driving electric cars will need to be designed responsibly today to reduce the pollution problems of tomorrow.*

To stay at 2018 decarbonization levels, the MIT study described challenges that we must begin to address now and suggested multiple responsible design approaches that could be considered.[5] For example,

- Algorithms could be designed to be more efficient without sacrificing safety, effectively reducing computing power requirements.
- New, less-polluting hardware designs could be developed that operate more efficiently to reduce the operational and computational workloads of self-driving cars. These could include more specialized hardware, ideally "future-proofed" to accommodate algorithmic improvements.
- Metrics could be established to assess the energy consumption impact of potential computational designs (i.e., you can't compare/improve what you don't measure).
- Industry standards could be created that require companies to release information about energy and computer power consumption for self-driving cars. These would provide critical data needed to assess and compare different designs and would set the stage for updated emission standards targeting AI-driven transportation.

The impact of these and other responsible tech design approaches? We will have more actionable strategies to reduce pollution and promote public health, and greater potential to reduce risk.

RESPONSIBLE DEVELOPMENT

Design provides a conceptual framework for processes, products, and systems. Development operationalizes design—translating the goals and specifications of the design plan to its realization.

How can we responsibly develop digital products and services that meet the aspirations of our designs? An effective approach: Test prototypes on real users to get a sense of how things will work. Iteratively pilot and improve the product or service in short-term sprints until it will work as intended for the customer and is ready for prime time. "Fail early and often" during development so the product will be less likely to fail drastically for users after its release.

This is the gist of an approach called *agile project development.*[6] Although agile development has its roots in a number of "lightweight," "lean," "adaptive management," and other software development methodologies, its launch is often attributed to the "Manifesto for Agile Software Development," created by a group of 17 leading software practitioners who met at a ski resort in Snowbird, Utah, in 2001.[7] The developers were frustrated by heavy-handed contracts that short-changed the user and the prevailing "waterfall" approach to software development that made it difficult to involve customers in product creation. In *waterfall development,* the product development cycle is linear with one phase being completed before the next: User requirements come before building, building is completed before testing, and so on. In agile development, building and testing are done at the same time and iteratively, user requirements can be modified throughout.[8]

The Agile Manifesto outlined a methodology based on four values:[9]

- *Individuals and interactions* over processes and tools
- *Working software* over comprehensive documentation
- *Customer collaboration* over contract negotiation
- *Responding to change* over following a plan

The idea was to integrate the user's perspective through an increasingly effective set of build-and-test pilots. The agile approach has the potential to embed empathy and an awareness of public impact—*responsibility*—throughout the development process. Today, agile software development methods are used in combination with other methods or as a main focus by the majority of project managers.[10]

Responsible development calls for using the right tool for the right job and it is up to the developer to use the approach most likely to get the best results. Former Honda R&D engineer and UMass professor Emily Nutwell points out that waterfall grew out of manufacturing: "When building widgets that are a known quantity, you need the discipline of waterfall. You need budget targets, what materials to buy, how many people to hire, etc. But with software, it is completely different. Projects are more emergent. We learn as we add features and different users. Processes are tied to the problems they are addressing."[11] She points out that agile isn't necessarily better than waterfall, it just takes a different approach.

This perspective is reinforced in a *Harvard Business Review* article that also contends that agile and waterfall approaches are not an either/or. Hybrid approaches that utilize waterfall for well-defined (often physical) components, and agile for less certain (often digital) components or components that benefit from more customer involvement, are most responsible.[12]

The transition to agile for software development has been a heavier lift in the public sector. Government agencies are often risk-averse about product development because of a mountain of requirements and a culture that rewards following established processes and procedures over creative problem-solving, an environment compellingly described by Jennifer Pahlka in *Recoding America*.[13]

Pahlka speaks from experience. She was deputy chief technology officer in 2013–2014 and cofounder of the US Digital Service in the federal government. Before joining the feds, she founded and headed up Code for America and worked with local, state, and federal governments. She and her colleagues often experienced the public sector as "the good, the bad, and the ugly"—a sector whose mission is to promote public well-being (the good), where dedicated public servants were often hamstrung by a development environment that makes compliance with process a higher priority than creating an effective product (the bad), and where products and services are not as effective, useful, or usable by the public as they should be (the ugly). Their experience with food stamps provides an example.

Why Was It So Hard to Apply for Food Stamps in California?

You would think that a development priority for public assistance programs such as food stamps would be to ensure that they are easy to navigate for the people who need them. But not always. In the early 2010s, Code for America, "the technology world's equivalent of the Peace Corps or Teach for America," worked with the state of California to explore why only 57 percent of Californians eligible in 2011 for the Supplemental Nutrition Assistance Program (SNAP)—food stamps—were enrolled.[14] The answer surprised them.

Code for America graduating fellows Jake Solomon, Dave Guarino, and Alan Williams sought to understand what was involved in applying to the program. What were the roadblocks? They found a 212-question form that was often confusing, seemingly irrelevant (Did applicants own a burial plot?), and sometimes suspicious ("Have you or any member of your household ever been found guilty of trading SNAP benefits for guns, ammunition, or explosives after September 22, 1996?"[15]). SNAP applicants couldn't apply for benefits on a mobile phone. It could take an hour to fill out the form—time many did not have. For those without devices, it was hard to apply on public library computers that restricted use to 30 minutes in an attempt to make sure that everyone had a chance to use limited resources.

The group worked on potential remediations that would make the form easier and more broadly accessible. Williams went the furthest to investigate problems with the SNAP system—he forwent his salary and applied for food stamps to experience the program up close. He tried and failed to enroll in the program on his mobile phone and on the computer at the library—access to fast Wi-Fi, an up-to-date device, and time mattered. It was awkward to check the balance on his electronic benefit card—key information that determines whether you can buy groceries and what you can afford.[16]

All of the problems with the form, the process, and accessibility took time, and big and small roadblocks mounted up for eligible participants—participants who often gave up before receiving benefits. Williams and his teammates created an app—GetCalFresh—that was a simpler and more effective way to sign up for the SNAP program. They actively promoted it to welfare directors in counties across California.

Between 2017 and 2025, GetCalFresh helped 6.2 million people receive more than $12.8 billion in SNAP benefits. It supported 81 percent of California's online SNAP applications by 2022.[17] Moreover, increased uptake of SNAP produced other benefits: Students in SNAP households tended to achieve greater educational advancement, low-income SNAP families had lower health care costs, and every dollar spent in SNAP benefits generated $1.50 in the local economy.[18]

From 2021 to 2023, Code for America worked with California on the rollout of a permanent, multilingual, multi-benefit application portal, BenefitsCal.com. In 2025, GetCalFresh was integrated into a unified California Statewide Automated Welfare System (CalSAWS) which provides a range of public assistance programs via the new portal.[19] The principles that guided GetCalFresh—plain language, mobile-first design, streamlined document uploads, and other user-focused features—influenced the development of the BenefitsCal.com portal and impacted how service delivery happens at scale across the state.[20]

Responsible technology development, like the development of GetCalFresh and its successors, ensures that products, systems, and services work for a broad and diverse clientele. Along the way, responsible development can integrate mitigations for risks and biases, promote ease of use, privacy, and safety for the user, and tread lightly on the planet. Standing in the shoes of target users, especially when making the myriad of development decisions—how to make things simple for customers, which technologies to use, which features to turn on or off, how to integrate products within complex solution strategies—applies a public interest tech mindset. Adopting this mindset can go a long way toward improving outcomes and reducing risks for the technologies critical to societal well-being.

RESPONSIBLE USE

Responsibility isn't just about the design and development of products and services. It's also about the way we use them. The seatbelt in your car can save your life, but it won't work if you don't use it. Seatbelts are now a universal addition to commercial vehicles, thanks to Title 49 of the United States Code (chapter 301, Motor Safety Standard), which took effect in 1968.[21] All states except New Hampshire require its use and many states require its use in the back seat as well. For many of us, even my friends in New Hampshire, responsible vehicle use involves seatbelts, whether or not it is the law.

In cyberspace, responsible technology use often involves *two-factor authentication*—something else that won't work if you don't use it. Two-factor authentication is a way to verify your identity through two different means, typically something you know (e.g., a username and password) and something you have (e.g., a smartphone). You often have the option (and are sometimes required) to do this online: You log in to your account (you know your username and password) and then verify it's you with a code sent to your smartphone or email or with your face or fingerprint (something you have). This approach is also used in the physical world: You get cash from your ATM by providing a card (something you have) and a PIN (something you know).

In 2010, the popular use and incorporation of two-factor authentication was spurred in part by Google's response to the hacking of the Gmail accounts of Chinese, American, and European human rights activists, and the exposure of private images (including nude photos of many celebrities) in Apple's iCloud.[22] These high-profile incidents alerted many to just how easy it was to access private information and how dangerous it could be.

Perhaps the most detailed and poignant recounting of the hacking of a personal account in the absence of two-factor authentication was by Mat Honan, a journalist for *Wired* magazine. Honan was hacked through a daisy-chained set of personal accounts (each account was linked to the next) by hackers who wreaked havoc on his digital life. He lost photos of his daughter, had his computers wiped of data, and much more: "In the space of one hour, my entire digital life was destroyed. First my Google account was taken over, then deleted. Next my Twitter account was compromised, and used as a platform to broadcast racist and homophobic messages. And worst of all, my AppleID account was broken into, and my hackers used it to remotely erase all of the data on my iPhone, iPad, and MacBook."[23] Using two-factor authentication—available on his Gmail account but not yet available on Amazon and Apple—would have limited the damage, as the hackers' real target was Honan's Twitter account.

Following these high-profile incidents and with the rising prevalence of smartphones, Google and Apple (as well as Dropbox, Slack, Twitter (now X), and a host of other companies) incorporated two-factor authentication as an extra layer of protection for user accounts in the 2010s.[24] Today, two-factor authentication is ubiquitous, although often still voluntary. Take a break from reading *now* and turn on two-factor authentication for your accounts.

Hacking incidents are still common, but hacking is harder than if we did not use two-factor authentication. The companies that incorporate these methods and the users who use them are responsibly nudging cyberspace toward public privacy, safety, and security.

RESPONSIBLE OPERATION

- In 2022, the Fern Hollow Bridge in Pittsburgh, Pennsylvania, collapsed, injuring 10 people, four seriously. Drainage problems had weakened and corroded the metal legs of the bridge.[25]

- Creosote builds up in the flue of your chimney, so much so that it can cause a chimney fire that may spread to your roof.
- You haven't updated or backed up your laptop in months. You click on a link that has a virus that erases your hard drive, causing you to lose critical data you can't recreate or resurrect.

What do these three things have in common? *Deferred maintenance.* The risks and damages of each might have been avoided had they been inspected or updated and maintained regularly. Being vigilant and regular about maintenance is critical for responsible operation of the physical and technological infrastructure on which we depend.

The American Society of Civil Engineers (ASCE) regularly grades America's infrastructure. It gives grades to infrastructure in 18 categories—from dams to drinking water to transit to schools. The grades reflect the health, safety, and risk of infrastructure. In 2025, ASCE gave America's infrastructure an overall C, with 9 categories receiving Ds or D+s. This was "good" news, as the infrastructure grade in 2021 was a C– with 11 categories receiving a D+, D, or D–.[26] ASCE attributed improvements since the last assessment to "the government and private sector prioritizing investments in systems that historically had received little attention."[27] From transit failures to bridge collapses to failing heating and cooling systems in schools, one thing is clear: Maintaining infrastructure and keeping it in good shape is crucial (and comes with a cost), but fixing or replacing broken infrastructure can have *catastrophic* costs—financial and otherwise.

Technology requires maintenance. After an IT product or system launch, maintenance provides a means of improving the software, boosting performance or security, correcting issues or bugs, and/or adapting software to new policies, regulations, or changes in other systems the software interacts with. Without maintenance, systems and products can become more vulnerable to hacking, harder to use with other systems, prone to data loss and system malfunction, and potentially unable to incorporate new functionalities.

Part of the problem with prioritizing maintenance is that it's not exciting, certainly not like innovation. In *The Innovation Delusion,* authors Lee Vinsel and Andrew Russell make the compelling case that maintainers and maintenance (often thought of as "dirty work" in service trades) should be prioritized alongside entrepreneurship and innovation.[28] They argue that promoting innovation and disruptive technology without prioritizing the maintenance that will ensure that innovation works and provides predicted benefits ultimately makes society "poorer, less safe, and—ironically—less innovative."[29]

This makes sense. Few people are enthusiastic about upkeep, and maintenance (technology and otherwise) is easy to ignore: It's inconvenient. Nothing bad has happened yet. You don't have time to update (and possibly reset the system to your liking). It may cost money you hadn't factored into the budget. But both the physical and the technological world demonstrate that regular maintenance—arguably the most low-risk, high-reward thing you can do—can help prevent catastrophe and promote the safety of all of us (and our data).

The damage that happens when individuals don't maintain their systems—a laptop crash, chimney fire, or vehicle brake failure—can be personally devastating. We are responsible for maintaining our personal infrastructure. But maintenance is even more critical for the organizations and agencies responsible for the infrastructure on which society depends—bridges, dams, the health care system. Maintenance must be built into *institutional* processes and budgets. Problems must be fixed.

And in many cases, this is what happens. Many companies proactively maintain software systems regularly: Apple, Microsoft, your car's computers, and other systems send regular updates that include security patches and fixes. When users ignore these updates, or when the company doesn't send them, we become more vulnerable.

How do organizations build in maintenance? They hire individuals—*maintainers*—whose job it is to keep things in good working order. They build sensors, systems, and monitors into products and services that give a heads-up about infrastructure health. They institutionalize maintenance through processes and procedures. These approaches work synergistically and often invisibly to make sure that infrastructure is in good working order and the risk is low. They provide a generous ounce of prevention to offset the dangers of many and costly pounds of cure.

Maintenance is a critical component of a responsible approach to infrastructure. As Vinsel and Russell write, "What would the world look like if we focused more on fixing things instead of throwing them away; if we learned to rely on and respect the everyday technologies that we take for granted rather than worship the new? . . . a maintenance mindset can lead to cultural and emotional well-being, and yes, even economic prosperity."[30] If we want to make technology responsibility mainstream, we need to build in maintenance—responsible operation.

MAKING RESPONSIBILITY STICK

The responsible design, development, use, and operation of technology is a great start. Full stop. We've gone a long way down the road toward nudging cyberspace to be good for humans. Is there more we can do to ensure that our efforts are most effective?

The best way to make change is to make your efforts habitual—to institutionalize good behavior. Anyone who's ever made a New Year's resolution to exercise knows this. You start off strong, but are you still exercising in March? April? If so, you are much more likely to reap the benefits of your exercise program—now integrated into your regular life—than if it was a one-off.

The same goes for the responsible design, development, use, and operation of tech. If these strategies are *institutionalized*—a normal part of what you do and embedded in organizational processes and procedures, job descriptions, and performance analyses—they are much more likely to change your and your organization's approach to tech and realize benefits. You will have regularized a public interest tech mindset.

For individuals, there are many ways to regularize our responsible use of tech. We do it when we consistently use two-factor authentication (making our accounts more secure) or deny nonessential cookies (protecting our privacy) when accessing a website. We do it by buying the more secure product (connected baby monitor, smart home appliance, etc.) and changing the password from the default (beefing up security). We do it when we proactively turn off unwanted location services or opt out of applications or services that inappropriately exploit user data.

Individuals can do a lot, but in the end there's only so much one person can do. We will go a lot further a lot quicker if we couple individual efforts with organizational efforts to increase tech responsibility.

Institutionalizing tech responsibility in organizational processes—the ones that guide how teams make decisions and what they prioritize—is critical. Tech responsibility must be part of an organization's goals, objectives, and key performance indicators. Without this, it becomes a "nice to have" and is often buried under more pressing priorities, pushed aside to meet deadlines, or skewed by the political dynamics and emotions in the room.

In recognition of the reach and risk of technology for humanity, no less than the United Nations has begun to promote and provide guidance for the responsible design, development, use, and operation of technology. The reason? "Responsible technology practices . . . are a manifestation of the UN's commitment to human rights, peace, equality, sustainability, and more."[31] In short, responsible technology practices are needed to promote the best interests of publics everywhere.

The UN goes further—from intention to follow-through. The *Responsible Technology Playbook*, developed in partnership with the technology consultancy Thoughtworks, provides a pragmatic and strategic approach that can help institutionalize responsible tech.[32] It references a cache of responsible tech tools, guidelines, and services that can help organizations and companies guide the tech they develop and that their customers use. The key takeaways from their pragmatic report: tech responsibility is promoted when individuals and organizations adopt a responsible tech perspective (i.e., act as public interest technologists), and responsible practice must be institutionalized and incorporated into design, development, implementation, management, and use of technologies—that is, at every stage.

The following seven strategies provide a framework for adopting a responsible tech mindset and institutionalizing responsible tech organizational infrastructure:

1. *Optimize for users.* Embed user involvement and feedback into organizational processes and procedures to drive product and service creation with the goal of meeting user needs and protecting their interests.

   Approaches like agile development (used by Code for America for GetCalFresh) institutionalize iterative improvement and regular user feedback. Incorporating such approaches that "learn from experts"—the users—ensures that products and services are more likely to meet user needs.
2. *Optimize for the environment.* Prioritize and incorporate strategies to develop products and systems that reduce power consumption, use materials and resources in an environmentally responsible way, and promote sustainability.

   Consider self-driving cars. They will have critical impacts on the environment and sustainability efforts due to the operation of their onboard computers and AI systems. They also rely on rare earth, lithium, cobalt, and other materials for batteries, some of which are in limited global supply. Planning for their environmental impacts must be considered as part of the responsible design, development, deployment, and resourcing of future vehicles.
3. *Build products and services for everyone.* Institutionalize design, development, and deployment strategies that result in products and services that are low-risk, accessible, useful, and usable by everyone—people from different cultural and age groups, people with different physical characteristics and abilities, and so on.

   We saw in chapter 2 that studies showed that facial recognition systems can be biased—less accurate at identifying some groups than others. Organizations can use auditing tools and other approaches to reveal biases in technology and provide workarounds to ensure that applications incorporating facial recognition systems minimize inaccuracies. In addition, services and products can, and should, be evaluated for broad accessibility prior to deployment, and monitored throughout public use.

4. *Create responsible solutions that fit the problem.* Use techniques that will protect the user: Avoid the tracking and collecting of unnecessary data (i.e., data not needed for product function) and the sharing of collected data with third parties not needed for its primary purpose. Customize AI to minimize errors and mitigate risk for all target users.

   Currently, "smart" cars collect a lot of data, some of which is personal (location; occupant internet and entertainment use; driver, passenger, and bystander images; etc.). Much of this data is needed to operate the car safely; it could also be sold, shared with partners, or exchanged at the discretion of the manufacturer. Taking a responsible approach to data means that personal data collected to operate the vehicle will only be used for this purpose and barred from being shared or exchanged otherwise.

5. *Be clear about risk and ethical challenges.* Regularly consider as part of institutional processes and procedures the broader impacts of products and services and the potential for intentional or unintentional misuse. Incorporate mitigations to reduce those risks. Ensure that the development and deployment of technologies meet the best ethical standards of the industry and community in which they will be used.

   Ethics isn't just the "right thing to do"; it can be actuated as a risk management strategy, a way of reducing risk to an organization's reputation, risk to the public, and often regulatory risk. Guidelines such as the UN's "Principles for the Ethical Use of AI in the United Nations System" can help drive ethical development and design.[33] Threat modeling can help identify potential risks. There are an increasing number of tools to help assess risk and guide ethical development and deployment.

   Concepts such as "ethics" and "risk" are somewhat fuzzy and hard to make concrete (as is "responsibility"), but the effort to assess and mitigate risk and to promote ethical development and use provides an ounce of prevention and helps reduce the need for a pound of cure when things go wrong.

6. *Promote success by incorporating means for measurement and accountability.* Include means for measuring and monitoring product and service usage and outcomes to ensure that public interest goals are being met. Make it an institutional priority to fix products and services when they are not meeting expectations.

   The business management guru Peter Drucker once said, "You can't manage what you don't measure." That applies to public interest goals as well. Responsible tech incorporates means to measure the success of products and services against public interest goals and improve them. That may involve the use of open-source techniques, which increase transparency and the pool of developers who can improve software.

7. *Promote success and drive responsible practice through organizational advocacy at every level of leadership.* Ensure that there are responsible tech advocates and champions in all development and stakeholder groups to drive processes. Give them the power, clout, and resources to achieve responsible tech goals.

These seven strategies can help realize a responsible approach to tech creation and use. They demonstrate that intention, institutionalization, and follow-through matter a lot, and if responsibility is made a priority, it can go a long way toward making cyberspace more beneficial and less risky for us all.

Responsible design, development, use, operation, and institutionalization all provide voluntary nudges, ways to proactively apply a public interest mindset to our engagement with cyberspace.

In the next chapter we consider "involuntary" nudges—laws, regulations, and policies that "make" us be responsible. Together, "voluntary" and "involuntary" efforts are needed to create an ecosystem where responsibility is *embedded* in the way we create, use, and manage technology, providing multiple ways to promote the benefits of cyberspace and reduce its risks.

4

RULES OF THE ROAD: LAW, REGULATION, AND POLICY

Congress probably never imagined that copyright law would change the direction of hip-hop, but it did. Hip-hop is a "musical mosaic" where collaboration between different artists, genres, and cultures is commonplace.[1] Using sampling—the inclusion of multiple snippets of music, film excerpts, speeches, and other media fragments—hip-hop artists in the '80s and beyond created deeply nuanced and innovative musical collages—both by working directly with collaborating artists and through incorporation of small samples of the work of others.

Do hip-hop creators need permission from the artists they sample? Sometimes there can be hundreds of samples, some so small that permission hardly seems worth it. But copyright law says otherwise and increasing application of US copyright protections in the 1980s, 1990s, and beyond changed the creative face of hip-hop.

The US Constitution grants Congress power to protect artists' exclusive rights to their creative work through copyright. This means that original creators must give their permission for others to use their work except in cases of criticism, commentary, parody, or creating transformative art. Was hip-hop any of these? That was the focus of an increasing number of hip-hop legal cases to determine whether the original artists were getting the credit and compensation the Constitution granted them.

The increased scrutiny of intellectual property in sampled works changed hip-hop. Public Enemy, Dr. Dre, Eminem, and other hip-hop artists changed their approaches largely because a genre born of mash-ups, free mixing, and remixing content from many artists was increasingly held responsible to the artists whose work was being used, even if there were hundreds of them in an individual piece. Said Chuck D from Public Enemy, "Public Enemy was affected because it is too expensive to defend against a claim. So we had to change our whole style . . . It's easier to sample a [single] groove than it is to create a whole new collage. That entire collage element is out the window."[2]

Law changes behavior. It impacts the way we engage with one another and the way we operate. In the case of hip-hop, it changed a musical genre. The use of law and policy as societal interventions has changed many other things: The 1963 and 1970 Clean Air Acts elevated awareness of public health and the importance of reducing air pollution and changed the way the auto and manufacturing industries operate. The Medicare and Medicaid Act of 1965 provided health insurance for citizens considered "bad risks" by private insurers—elderly and low-income populations.[3] Many decades later, we are still working through these environmental and public health challenges, but laws and policies have served as important interventions needed to make real progress.

The same is true for cyberspace. Law, regulation, and policy have the potential to change the way we manage cyberspace and to reduce its risks. They serve as guardrails for the public. But developing effective guardrails is challenging. Technology-relevant law, regulation, and policy must promote beneficial outcomes for humans impacted by complex and dynamic systems—systems with unanticipated consequences, systems whose "actors" may be algorithms or AI agents not subject to penalties in the same way humans are, systems for which there may be no current legal framework for assigning responsibility and liability.

In this chapter, we focus on the challenges of creating laws and policies that promote human well-being in cyberspace. We highlight three key areas with legal and policy innovations—digital rights and the General Data Protection Regulation, AI risk and the AI Act, and management of the internet and net neutrality policy. In chapter 5, we will continue our focus on law, regulation, and policy and explore what it takes to ensure that these societal interventions can effectively achieve their public-focused benefits.

Looking ahead, advances in AI are moving us toward a hybrid environment where decisions are made by both people and machines. Future scenarios will need governance that incorporates the increasing impact of AI decision-making in a way that promotes the interests of humans, does not diminish their agency or exacerbate their risks, and holds bad actors accountable when things go wrong. We consider what it will take to develop law and policy for this hybrid environment in chapter 8.

For now, we focus on the present. Before we get started, we briefly review key tools in our intervention toolbox—law, regulation, and policy.

WHAT'S THE DIFFERENCE BETWEEN LAW, REGULATION, AND POLICY AND WHY DOES IT MATTER?

"There oughta be a law" is a common sentiment when people think the public sector should be setting the ground rules. Although we don't usually say "there oughta be a regulation" or "there oughta be a policy," what we commonly mean is that there should be some proactive intervention that protects the public. Whether that intervention is a law, regulation, or policy changes the way it is created and its potential impact and outcomes. Knowing the tools in your toolbox is important to effect change.

In the United States, (federal) *laws* are instruments of the legislative branch and are passed by Congress and signed by the president. They are largely used to establish requirements and prohibitions. They are binding and ideally effectively enforced. (Federal) *regulations* are published by executive branch agencies to clarify the interpretation of a law and how it will be implemented. Regulations also state requirements or prohibitions and hold the force and effect of law. *Policies* establish guidelines or principles that shape decision-making and define desirable normative behavior and operation. They are nonbinding and lack the legal force of law and regulation.

The creation of laws is a complex process that involves a lot of horse-trading. Political parties, lobbyists, experts, and other stakeholders have a strong influence on the language and conceptualization of laws. Individuals and groups can impact the process through efforts to engage with the legislators who create and/or will vote on the law, and through petitions, referendums, public advocacy (including social media), and participation in hearings. Strong public opinion can make a difference. For example, the Civil Rights Act of 1964 was propelled by sustained advocacy by citizen groups and others.

The development of regulations provides a more open avenue for public influence. Many regulatory agencies are required to include a public comment period that gives citizens and groups an opportunity to provide feedback on proposed regulations. They may also create advisory committees that include members of the public. Public comments can make a difference. The Environmental Protection Agency (EPA) and the US Army Corps of Engineers' Clean Water Rule elicited over a million public comments.[4] It's not just the feds. Local and state governments also pass laws and regulations applicable to their jurisdictions.

Policy is often the first step before law and regulation. Policies are created by both private sector and public sector organizations and by local, state, and federal governments. They are legally nonbinding but, when effective, drive desirable normative behavior. Citizens have a variety of tools to shape policy: advocacy, protests, social media, community organizing—these can work for influencing law and regulation as well. Strong citizen sentiment can make a big difference. In 2014, the Federal Communications Commission received an unprecedented 3.7 million comments on changes to the Open Internet Order (net neutrality), many of which were driven by an informative and hilarious episode on the rule by John Oliver on *Last Week Tonight*.[5] (More on this in chapter 6.)

Whether the intervention is law, policy, or regulation, citizens and groups can make a difference. It's important to use all areas of influence available to promote the public interest—as do stakeholders with competing interests. In this chapter, we discuss three key legal and policy interventions that have led to the creation of public-focused law and policy in cyberspace. We begin with a pioneer in digital rights legislation—the European Union's General Data Protection Regulation.

LEGISLATING DIGITAL RIGHTS: THE EU'S GENERAL DATA PROTECTION REGULATION

Since the commercialization of the internet and its widespread use by consumers, data has arguably been cyberspace's most precious commodity. As digital technologies have increasingly become critical infrastructure, concerns about the privacy of personal digital data—your preferences, opinions, contact information, demographics, sexual orientation, and so on—have increased. This information is regularly bought, sold, exchanged, used for purposes other than provided, and sometimes made public without permission. Governments everywhere and at all levels are working to address the challenges of determining just what an individual's digital rights should be, and how to develop proactive interventions that incorporate protections for private and personal data.

As we'll see in chapter 5, digital rights and privacy legislation vary widely around the world. That being said, the approach taken to digital rights over the last few decades in Europe has been considered a global model. The influence of the EU's digital rights legislation, and especially the General Data Protection Regulation (GDPR), has been widespread. Its comprehensive legal approach and inclusion of effective penalties have provided a structure for digital rights management that positively impacts individuals, puts organizations on notice, and has influenced other nations and jurisdictions. As stated in a comprehensive analysis of the GDPR, "The GDPR is the most consequential regulatory development in information policy in a generation. The GDPR brings personal data into a detailed regulatory regime, that will influence personal data usage worldwide."[6]

The General Data Protection Regulation was adopted in 2016 and began to be enforced in 2018. It built on 20 years of broad consultation, analysis, and policy concerning the digital rights of EU citizens and an earlier Data Protection Directive. In 2025, the EU continued to evolve its approach to digital privacy by considering simplifications to parts of the GDPR to make reporting requirements for small businesses less onerous, include allowances for AI, and refine use cases and exemptions for personal data.[7]

The GDPR provides a comprehensive structure that specifies principles, infrastructure, and accountability mechanisms to protect the security and privacy of the data of EU citizens. It aligns with Europe's deep constitutional commitment to personal data protection, considered to be a reconfirmation of human rights "in the electronic dimension" and an extension of "respect for the human body."[8] Importantly, the GDPR applies to any company that processes the data or offers goods and services to Europeans, *even if the company is not based in the EU.*

The GDPR is comprehensive legislation that includes the following elements applicable to personal data within its scope:[9]

- *Data protection principles* that create a framework for governance. Data must be collected transparently, processed for purposes specified explicitly to the data subject and only for as long as needed, and handled to promote appropriate security, integrity, and confidentiality of personal data.

- *Individual data privacy rights.* These rights include the right to be informed and to access, rectify, and erase your data, the right to move your data from one company to another, and the right to restrict processing of your data.

- Data is primarily used only with the *data subject's consent* and for stated purposes. In some cases, the GDPR does allow data to be processed without permission—to comply with a legal request, to save someone's life, or to carry out an official function. In these instances, its use must be transparently shared with the subject.

- *Data privacy and security* must be supported "by design and default." New products and services must incorporate the GDPR's data protection principles, and security breaches must be shared with data subjects within 72 hours.

- *Accountability* is promoted by the appointment of data controllers and data protection officers within organizations, by overarching data protection authorities, by required documentation of data collection and use, and by substantial penalties for lack of GDPR compliance.

The GDPR's clear prioritization of individual data rights and privacy, facilitated by an explicit legal support and accountability structure, was groundbreaking. It's even more impressive if you consider the task of legislating within the European Union. Imagine if in the United States each state had its own language, culture, and currency, and North Dakota and South Dakota had different approaches to cross-border activities and treaties than Georgia and Florida. This is analogous to the geopolitical landscape in Europe before the long ramp up to the formation of the European Union. The EU had to create an effective confederation of countries whose histories and cultures were thousands of years in the making and distinct from one another. It is striking to realize that it was only in the 1990s that the EU introduced the euro as a uniform currency and embraced border-free travel and a single European market.[10]

As it approached the twenty-first century, the EU saw itself as lagging in terms of use and deployment of information and communication technologies. In 1995, the European Commission adopted a Data Protection Directive—a comprehensive approach to harmonize diverse national privacy laws and inconsistent enforcement of existing laws. The GDPR was developed to address experiences with the Data Protection Directive's shortcomings. Beginning with a broad set of expert consultations in 2009 and proactive consideration anticipating how information practices might be manipulated to evade regulatory goals, an initial proposal text of the GDPR was published in 2012. Agreement on a revised text was reached in 2015 and the GDPR was officially adopted in 2016 with enforcement starting in 2018.[11] Current updates to the GDPR continue this trajectory, factoring in present political dynamics and experience to date with the regulation and its implementation challenges.

Since 2018, the EU has continued to legislate to promote digital rights and to extend, strengthen, and complement the GDPR. Subsequent legislation has included the Digital Markets Act (regulating "gatekeeper" companies to provide fairer markets and more opportunities for competition), and the Digital Services Act (providing oversight of social media and other platforms with respect to content moderation, misinformation, and algorithmic transparency). Other complementary legislation targets cybersecurity, confidentiality in electronic communications (the "cookie law"), and resilience of financial data. In late 2025, the EU proposed a "Digital Omnibus" package that harmonizes and streamlines existing digital regulatory frameworks and seeks to be more "innovation-friendly" while remaining protection-focused—a sign of the times.[12] The text of the Digital Omnibus package will evolve as it moves through the EU legislative process. This aligns with the EU's general approach of adjusting and adapting its body of digital rights legislation to reflect national protection priorities, experience with legislative outcomes, and the dynamics of a changing world.

The GDPR has had far-reaching impact. It applies to EU citizens outside of Europe and non-EU companies operating within Europe. To address the inconsistency of privacy legislation around the world, many companies have adopted a single approach that is compliant with the more stringent GDPR, even though other countries may support weaker privacy protections—a business decision that supports both greater efficiency for companies and greater protections for their users.

Students often ask me: Can't the US just adopt the GDPR to promote digital rights and privacy in our country? The answer is no. The United States has a different governance structure and a different ideological and political approach to the rights of individuals versus the rights of corporations than the EU. Much of US federal privacy legislation is focused on specific groups or areas (children, health care, credit information) and much of our privacy legislation is enacted at the state level. There is no general right to privacy in the US Constitution.

Importantly, US privacy law primarily provides protections against the misuse of personal information and is focused on the *prevention of harm*. In contrast, the EU approach *protects rights*. The contrast between harm prevention and rights protection means that digital rights and privacy law in the United States will need to be different than in the EU.[13] Led by California, some states in the US are beginning to enact more comprehensive digital rights and privacy laws, which may push the US toward rights protection in addition to harm reduction.

In the United States, promotion of privacy and digital rights is also highly influenced by the administration in power, its approach to legislation, and its use of US agencies to enforce protections. For example, in the early 2020s, the US Federal Trade Commission stepped up its efforts to protect privacy by more strongly enforcing existing laws and levying fines on companies with insufficient privacy and security protections. With the transition of presidential administrations in 2025, some efforts were rolled back, and new efforts are expected to align with the current administration's priorities.

Ultimately, for the United States to create digital rights and privacy protections with the impact and reach of the GDPR, *it will be important to define just what general digital rights and rights to privacy Americans should have*. What would a US "Bill of Digital Rights" look like? Legislation will need to specify how those rights should be enforced and by whom, as well as identifying specific public sector agencies and infrastructure to ensure enforcement and accountability. Until then, the GDPR and other protective digitally focused legislation from the EU will continue to influence the rights and privacy landscape from afar, both for the United States and for many other countries around the world.

LAW AND POLICY TO REDUCE THE RISKS OF AI

As AI becomes more powerful and a more dominant force in cyberspace, perhaps the greatest public interest challenge is how we should develop, manage, use, and regulate AI to promote human well-being and minimize AI's risks.

This is tricky. AI is rapidly advancing, both through research into new functionality and capabilities and through the increasing incorporation of AI systems into the products, systems, and services of cyberspace. Any effective attempt to rein in the power and perils of AI can neither focus on flicking the OFF switch on research nor on barring a specific product or application—another one will rapidly take its place, perhaps even before legislation can be debated and passed.

Attempts to manage the risks of AI must focus on its outcomes, giving companies the latitude to ensure that safer outcomes are achieved in whatever way makes sense. This also ensures that as newer and more powerful products and services are developed, they must align with specified low-risk, public-focused outcomes that promote human well-being in cyberspace.

With the striking recognition of the power and peril of AI demonstrated by ChatGPT, Dall-E, and other AI commercial systems in the early 2020s, nations have sought to create legislation, regulation, and policy that would minimize the most serious risks of AI and protect the public. In the European Union, following more than a decade of serious discussion and experience with the GDPR and digital technologies, the AI Act was proposed in 2021 by the European Commission, and its specifics were approved in 2023 by the European Parliament. A landmark piece of legislation, the AI Act focuses on limiting the worst risks of AI to humans. This approach is both groundbreaking and eminently sensible: Focus on limiting the application of AI in the areas where AI can do the most harm. These areas include "health care, education, border surveillance, and public services, as well as banning uses that pose an 'unacceptable risk.'"[14]

The risk levels delineated in the EU AI Act vary in severity:[15]

- *Unacceptable risk*—AI systems considered to be a threat to humans. These systems are banned. They include systems used for social scoring and predictive policing, systems that manipulate cognitive behavior or specific vulnerable groups (e.g., children), and other systems with the potential for human harm.

- *High risk*—AI systems considered to negatively affect safety or fundamental (EU) rights. These include systems or products in critical infrastructure, in employment and hiring procedures, or that are governed by EU product safety legislation. These systems are required to incorporate human oversight, risk management infrastructure, and other protections.
- *Limited risk*—AI systems that pose specific but not severe risks. These include systems that manipulate image, video, or audio content and include deepfakes. These systems are subject to specific transparency standards to help the public make informed choices.
- *Minimal or no risk*—AI systems that have minimal risk impact. In 2021, these included video games and spam filters. Systems at this risk level are not restricted.
- *General purpose AI systems*—General purpose AI systems must comply with copyright and transparency requirements about training data. Those that present systemic risk (meet high computing power thresholds, have broad impact, or are considered to have elevated potential for harm) must conduct model evaluations, adversarial testing, track and report serious incidents, and ensure cybersecurity protections.

Remediations and penalties vary by the severity of risk, with the most scrutiny and restrictions and the highest fines coupled with the highest-level risk categories. In proposing different remediations for different risk levels, the EU AI Act is strategic, pioneering, and sets a model for the rest of the world. It imposes evaluation requirements and transparency requirements for generative AI systems. It sets up new organizational infrastructures to ensure implementation, compliance, and enforcement. It includes large penalties for lack of compliance. Like the digital rights–focused GDPR before it, the EU AI Act is poised to impact how companies, governments, and citizens in the EU and elsewhere develop and experience AI.[16] It goes into effect between 2025 and 2027 on a rolling basis with full compliance for high-risk AI systems expected by 2027.

Experience with these guardrails will no doubt lead to future adjustments, similar to the GDPR.

The AI Act was influenced in part by the 2019 recommendations from the Organisation for Economic Co-operation and Development (OECD), a consortium of 38 democracies with market-based economies that collaborate to develop policy standards that promote sustainable economic growth. The OECD AI recommendations focused on the use and development of trustworthy AI, specifying key target areas in the AI ecosystem:[17]

- Inclusive and sustainable growth and well-being
- Human-centered values and fairness
- Transparency and explainability
- Robustness and safety
- Accountability

The OECD included specific recommendations to invest in AI research and development, shape an enabling policy environment for AI, build capacity and prepare for workforce transition, and increase focus on trustworthiness in AI. The OECD recommendations have been influential in Europe, the United States, and other jurisdictions in their efforts to conceptualize a governance framework to manage the societal impacts of AI in cyberspace.

In the early 2020s, specific AI legislation in the United States was lagging, although the groundwork for AI legislation and policy was being laid. Influenced by the OECD recommendations, the US Office of Science and Technology Policy (OSTP) released a Blueprint for an AI Bill of Rights in 2022. The blueprint focused on five principles to "guide the design, use, and deployment of automated systems to protect the American public in the age of artificial intelligence." Together, these principles constituted a general framework for public-focused AI and tech guardrails:[18]

- *You should be protected from unsafe or ineffective systems.* Automated systems should be developed with broad consultation, tested before deployment, and monitored to identify and mitigate risk. They should be designed to proactively protect you from harms stemming from unintended, but foreseeable, uses or impacts. Independent evaluation and reporting should confirm that systems are safe and effective, and the results should be made public whenever possible.

- *You should not face discrimination by algorithms, and systems should be used and designed in an equitable way.* Creators and deployers of automated systems should take proactive measures to safeguard individuals and communities from unfair treatment or negative impacts based on protected characteristics such as religion, race, ethnicity, disability, or sex. Independent evaluations and algorithmic impact assessments should be performed and made public whenever possible to confirm these protections.
- *You should be protected from abusive data practices via built-in protections, and you should have agency over how data about you is used.* The public should be free from unchecked surveillance. Users should have meaningful control over how their data is collected, used, shared, and deleted. Privacy should be built into automated systems by default with data collection limited to what's necessary and aligned with reasonable expectations. Sensitive data—such as health, education, work, finance, or criminal justice information—should have stronger safeguards, ethical review, and limits on use.
- *You should know that an automated system is being used and understand how and why it contributes to the outcomes that impact you.* There should be plain language and transparency about when an automated system is being used, how, and why. Automated systems should provide explanations that are technically valid, meaningful, and useful to you, as well as calibrated to the level of risk based on the context.
- *You should be able to opt out, where appropriate, and have access to a person who can quickly consider and remedy problems you encounter.* Human consideration should be accessible, equitable, effective, maintained,

accompanied by appropriate operator training, and should not impose an unreasonable burden on the public.

The principles were intended to apply "to 1) automated systems that 2) have the potential to meaningfully impact the American public's rights, opportunities, or access to critical resources and services."[19] The seeds of protections for digital critical infrastructure had been planted. In 2023, the Biden administration issued Executive Order 14110 on the "Safe, Secure, and Trustworthy Development and Use of Artificial Intelligence" based on these principles, directing federal agencies to take concrete actions.[20]

The five principles of the OSTP's 2022 AI Bill of Rights and Biden's 2023 executive order represented the priorities of the administration in power. But priorities often change with presidential transitions. In 2025, the incoming Trump administration rescinded Biden's executive order on AI during their first week in office. Biden's executive order was replaced with an executive order entitled "Removing Barriers to American Leadership in Artificial Intelligence."[21] Trump's executive order realigned priorities to strengthen the US's AI advantage relative to other countries and increase the alignment between AI and the new administration's energy priorities. Protections were out, primacy was in.

The Trump administration put "meat on the bones" of the new executive order in July 2025 with the release of a new AI Action Plan.[22] The AI Action Plan focused on recommendations to accelerate AI innovation, build AI infrastructure, and lead in international AI diplomacy and security. The administration's priorities were clearly stated on the first page: "We need to build and maintain vast AI infrastructure and the energy to power it. To do that, we will continue to reject radical climate dogma and bureaucratic red tape, as the Administration has done since Inauguration Day. Simply put, we need to 'Build, Baby, Build!'"[23]

Some recommendations in the action plan focused on abandoning the "onerous regulatory regime" laid out by the previous administration and to "work with all Federal agencies to identify, revise, or repeal regulations, rules, memoranda, administrative orders, guidance documents, policy statements, and interagency agreements that unnecessarily hinder AI development or deployment."[24] The plan encouraged expedition of environmental permitting for modern AI data centers by "streamlining or reducing regulations promulgated under the Clean Air Act, the Clean Water Act, the Comprehensive Environmental Response, Compensation and Reliability Act, and other relevant related laws."[25] Some protections were promoted, especially for AI-driven critical infrastructure. The plan included recommendations to bolster cybersecurity and "secure-by-design" approaches for safety-critical or homeland security AI applications. It provided guidance to include AI systems in the establishment of standards and response frameworks, and within CISA's Incident & Vulnerability Response playbooks.

How all this will play out is unclear. Ideally, AI primacy (a national goal) and public protections (a public interest goal) should be able to coexist. Effective legal, regulatory, and policy protections are crucial to ensure that the outcomes of AI systems do not pose harmful risks to the populace. When those protections are deprioritized, society must resolve a deep and fundamental question: Can nations afford to endanger their publics and the planet to make sure they come in first in the AI race? Resolution of this question will play out before our eyes. We must not be bystanders.

POLICY AND MANAGEMENT OF THE INTERNET

Sometimes laws and regulations are not the right interventions for promoting the public interest. They may be too heavy-handed, too abstract, too formal, and not well-suited for specific environments. Common policies—"you must be this tall to ride the roller coaster" at Disneyland, "carry out what you carry in" at Yosemite National Park, or "no smoking" in many restaurants—are created to drive normative behavior that promotes safety, environmental sustainability, and/or public health. Sometimes they are backed up by law and regulation, but sometimes they just work to promote desirable behavior in a small or circumscribed environment.

At the other extreme, policy may also be the right tool to shape normative behavior in a complex, multijurisdictional environment when there is no one overarching authority. This is especially true of the internet, a decentralized, diverse, and global system where no one country calls the shots and stakeholders span users, providers, countries, multinational companies, and others.

Policy provides the governance structure for the internet. Its global community of stakeholders intend for it to be open, secure, and accessible while also promoting innovation, individual rights, and the interests of governments and stakeholders. This is a tall order. To create the worldwide buy-in needed for governance, internet policy and standards are developed by a broad set of international groups. They include the Internet Engineering Task Force (creating technical standards for internet operation and interoperability), the United Nations' International Telecommunication Union (focusing on global telecommunications standards, spectrum allocation, and other issues), the UN Internet Governance Forum (focusing on public policy issues such as internet accessibility and online safety), and others. Networking expert Jim Kurose's YouTube video "Who Controls the Internet?" provides an excellent description of the main players in internet policy and the complexity of their interactions.[26]

In addition to policies developed by international groups, countries have a big say in how the internet is managed within their jurisdictions. In the United States, one of the most important areas of policy has to do with whether the internet is managed as an essential service like a utility, in which all internet content and all users are treated equally (*net neutrality*), or as an information service, where some users can be treated preferentially (often called *pay-to-play*). Many see net neutrality as equity protection for access to the internet, its apps, services, and information.

Net neutrality policy prohibits *internet service providers* (ISPs, the organizations that deliver access, management, and other services needed to run the internet) from granting special access to specific customers or favoring specific traffic in the allocation of internet resources. In a pay-to-play model, differential treatment is permissible. For example, under a pay-to-play regime, Netflix, Hulu, or other streaming services could pay for a "fast lane" on the internet to ensure their content gets to your device faster than that of companies using a more conventional "slow lane."

The difference is critical. Treating the internet as a utility (net neutrality) essentially applies the assumptions about equitable access we use with nondigital critical infrastructure to the internet. The argument goes like this: If the internet is considered a utility and critical infrastructure, it should be used and managed in a nondiscriminatory way. We expect equal opportunity and access to other critical infrastructure (water, power, emergency services, etc.)—anyone can buy *more*, but you can't buy *better*. Net neutrality would bar internet service providers from providing different levels of service to different classes of people at the discretion of the provider. It may also subject ISPs to additional regulation and policy restrictions.

With pay-to-play, the internet is less likely to be subject to regulation and becomes a freer environment for innovation. Those against pay-to-play argue that the fast lane (for those who pay) will be preferred, making the slow lane feel like the dial-up connections of the 1990s, and that internet companies will have too much power. Pay-to-play advocates claim that service will be faster, network quality will be better, and that pricing is more likely to be commercially reasonable.[27]

Note that the architecture of the internet gives ISPs the ability to provide different priorities to different data packets transmitted on the internet and to block packets as they choose. Under *either* net neutrality or pay-to-play, this functionality must be retained for the smooth operation of the internet: Some protocols have precedence over others and some packets, like botnets, will need to be blocked. But the ability to prioritize packets also provides a medium for granting different levels of service based on ability to pay, preferred partners, or specific content, all of which net neutrality advocates feel should not be the reason for prioritizing some internet users and customers over others.[28]

In 2015, the US Federal Communications Commission specified three clear "bright line[s]" to define the outcomes we would expect with net neutrality:[29]

- "*No blocking.*" This bars ISPs from blocking lawful content, applications, and services. Your ISP can't decide you are especially worthy or unworthy based on their own criteria (subject to reasonable network management).
- "*No throttling.*" This bars ISPs from impairing or degrading (slowing or delaying) internet content, applications, or traffic at their discretion (subject to reasonable network management).
- "*No paid prioritization.*" This bars ISPs from favoring some internet traffic over other internet traffic based on the sender or receiver's ability to pay (subject to reasonable network management).

Although these outcomes provide clear goals that promote equal opportunity and access to the internet, their interpretation and the details of their enforcement are challenging: What constitutes "reasonable network management"? Sometimes dropping or prioritizing internet traffic is a technical necessity (e.g., during network overload). It can also be a business decision (e.g., prioritizing an ISP's own service over a competitor's). In a pro-net neutrality environment, how should this policy be effectively implemented and enforced?

When I began to write this book, net neutrality was a policy supported by the federal government; today it is no longer upheld. The shift in government support of net neutrality varies regularly in the United States because it is a nonbinding policy—not a law—and enforced or not depending on the preferences of the administration in power. The many stakeholders in cyberspace, including companies, lobbyists, users, and political parties, see net neutrality's value and limitations from different perspectives, and often as a conflict between internet equity of access and opportunity versus the freedom to innovate and to promote a healthy (capitalist) internet economy. Because of this, the FCC regularly flip flops with respect to its support for net neutrality. In the absence of enforceable law and regulation, the application of net neutrality policy is inconsistent.

If or when net neutrality becomes law, navigating its practical challenges and the development of an effective structure for enforcement and accountability will be critical. Regulations will need to be developed to determine what constitutes reasonable network management and what constitutes preferential treatment, and one or more agencies or groups will need to be designated for effective enforcement. The development of law and regulation may be even more challenging than the development of policy, but as Kurose points out (quoting Tim Wu[30]), "'when invention begets industry and industry begets empire,' the stakes couldn't be higher."

GUARDRAILS FOR A DYNAMICALLY CHANGING LANDSCAPE

The GDPR, the AI Act, and net neutrality are large-scale societal interventions focused on the interests of large publics. In each case, years of discussion, debate, and engagement with many groups and all sectors preceded the law, regulation, or policy.

But societal interventions can happen for all kinds of tech and at every scale. Scroll your news app (or "open the paper"), and there are many instances of new laws and policies being developed to protect the public from the risks of new technology capabilities or unhealthy use. These instances constitute proactive interventions to rein in risk and are perhaps the seeds of next-generation legislation. It is important that they be developed thoughtfully, and with a public interest tech mindset, as their enforcement will bring both anticipated and unknown outcomes.

Consider the schools in your city or town. Many are grappling with two current tech-related challenges: the education-related risks of excessive student cellphone use, and the availability of AI assistants. Both challenges perturb the role of instructors and the educational experience of students (often not in a good way). To address them, we are seeing the creation of new policies that seek to control student use of cellphones and personal devices in the classroom, as well as the use of AI assistants for schoolwork, both for the purpose of guiding better learning outcomes for today's students.

Take student cellphone use. As we saw in chapter 1, teenagers spend a lot of time on their phones, including in school, and educators are concerned. Building on over a decade of research on student cellphone use, addiction to social media, and the impacts of technology use on academic performance and kids' mental health, many schools are beginning to formally restrict or prohibit the use of cellphones, smartwatches, and other personal devices on campus. A 2024 National Education Association (NEA) poll indicated that educators support restrictions, reporting that 90 percent of its members approve of school policies prohibiting student use of cellphones and personal devices during instructional time.[31]

As of summer 2025, at least 33 states and the District of Columbia had passed laws or created policies to require school districts to ban or restrict students' use of wireless personal communication devices in school. Florida, Louisiana, South Carolina, and Utah went the furthest with a *statewide* restriction barring students from using these devices unless permitted by teachers for educational purposes (with exemptions for students who need their devices for medical reasons or as part of an individualized education plan). South Carolina made the implementation of a cellphone policy mandatory for schools receiving state funding.[32]

On the ground, teachers continue to worry that student use of smartphones, smartwatches, and other devices takes a serious toll on their ability to focus in the classroom. Today's new policies about cellphone use in schools are intended to be proactive interventions to rein in tech overuse, addiction, and distraction, and to better promote student learning. It will be important to track educators' experiences with these restrictions and to evaluate their impact on classroom culture and learning. Based on what we learn, we may need to fine-tune these policies to ensure that they are meeting the tech-related educational challenges—and opportunities—of today's world.

The challenge of AI assistants in the classroom is more recent and different—less about distraction and mental health and more about teaching and learning. The increasing prevalence and capability of AI assistants are both an opportunity and a challenge. AI chatbots can be used to enhance students' writing and research skills. (See the appendix for details on how I use AI in my own classes). AI tools are also being used by teachers to complement instruction on everything from the arts to medical science.

On the other hand, AI can be a crutch, and students faced with busy schedules or a lack of motivation may use it as a shortcut on homework designed to build their own skills. The lack of skills—writing, coding, assessment, critical thinking—makes it harder to use AI productively and meaningfully as students whose skills are weak are not well equipped to evaluate or verify the output of their AI chatbots. The challenge for many educators is to design curricula that both builds student skills *and* fosters awareness of how to productively and safely use the AI tools that students will encounter in the workplace.

To this end, many universities are now adopting institutional policies about the use of AI in the classroom, and more of these policies focus on responsible use rather than outright prohibition. In a 2024 policy statement, the NEA emphasized that AI should be used to enhance the educational experience and that "humans must always be the center of the teaching and learning experience and play a significant role in every consequential education and employment decision."[33] As always, the devil is in the details.

As teachers and students continue to experience both the positive and negative impacts of AI in the classroom, today's emergent institutional AI policies are providing a variety of examples of how to deal with AI in education. At many institutions, including my own, we are beginning to see policies that couple guidelines about AI use with training on AI literacy and safety. This approach also provides an important opportunity to help students develop a public interest tech mindset about the responsible use of today's technologies, both in school and for future employment.

Beyond the United States, countries all over the world are grappling with how to incorporate AI in education and expand AI literacy. In China (as intent on the race for leadership in AI as the US is), AI is being integrated into the national curriculum as a compulsory subject for primary and secondary school students. South Korea is investing in digital infrastructure for teacher training. Australia created a national Framework for Generative AI in Schools in 2023, as has the United Arab Emirates.[34] These efforts indicate the importance of policy guardrails and investment for publics everywhere to manage the ubiquity and power of the technologies that have become essential to society and are becoming essential in education.

WHAT MAKES LAW, REGULATION, AND POLICY EFFECTIVE?

Any gardener will tell you that you can't get the most out of your plants if you start with bad soil. This applies to law, regulation, and policy as well. They are not effective if they don't make sense. They are also not effective if they are not enforced, or if the penalties for noncompliance are inconsequential. To create effective law, regulation, and policy, we need to give them the best possible chance to succeed when we develop them and the best possible chance to make a positive difference once they are operational.

How do we "prepare the soil" to create effective law, regulation, and policy? Expertise is needed to create realistic projections of societal outcomes—how potential conceptualizations of law, regulation, and policy will reduce risk and protect people. Enforcement mechanisms, incentives, and/or penalties must be incorporated into the final language to increase the effectiveness of these societal interventions and create meaningful consequences for noncompliance. Moreover, all interventions are most effective if they are well-aligned with the political philosophy and governance model of the jurisdiction in which they will be applied. In the next chapter, we focus on how creating the right context for the development and deployment of effective law, regulation, and policy—preparing the soil—can help achieve their potential to promote positive change.

5

NECESSARY BUT NOT SUFFICIENT: HOW DO WE MAKE PUBLIC-FOCUSED LAW, REGULATION, AND POLICY EFFECTIVE?

After reading chapter 4, the skeptical among us will point out that there are lots of laws, regulations, and policies that people ignore—interventions that don't have impact because they are not enforced or because the penalties are too low for compliance. Most people I know (including me) jaywalk. Lots of people litter. In most places, you can go over the speed limit a little without getting a ticket.

When we ignore laws, regulations, and policies, the change they were created to foster doesn't happen. In this chapter, we focus on what we can do to increase their effectiveness and drive the public interest outcomes they were developed to promote. In particular, we consider three aspects of the effective development and deployment of these societal interventions: *expertise* needed during their creation to promote the outcomes they are developed to drive, *enforcement* needed during their deployment to ensure compliance, and *alignment* needed to ensure that societal interventions are reinforced by the prevailing culture, style of governance, and oversight in their jurisdiction.

We begin with a discussion of expertise and an iconic moment that demonstrated just how little many US elected officials know about tech.

CHAPTER 5

EXPERTISE: CREATING EFFECTIVE LAWS AND POLICY IN THE FIRST PLACE

In April 2018, Mark Zuckerberg, CEO of Facebook (now Meta), appeared before the US Senate's Commerce and Judiciary committees to testify about data privacy and Russian disinformation.

The hearing was held in the wake of the Cambridge Analytica scandal, a controversy in which a research survey of Facebook users—"This is Your Digital Life"—was leveraged by the UK company Cambridge Analytica to collect data on up to 87 million Americans. The information gathered was subsequently utilized to assist the presidential campaigns of Ted Cruz and Donald Trump in their efforts to influence undecided voters through advertisements tailored to each user's emotional and psychological triggers (also known as psychographic targeting).

What was striking about the testimony at the hearing was not what Zuckerberg said but the general lack of understanding the Senators had about what Facebook does and how social media actually works. Senator Orrin Hatch from Utah asked Zuckerberg, "How do you sustain a business model in which users don't pay for your service?" (After blinking for a moment, Zuckerberg responded "Senator, we run ads.")[1] Senator Brian Schatz from Hawaii asked Zuckerberg whether Facebook could see the emails he sends on WhatsApp, owned by Facebook.[2] (You send messages via WhatsApp and they are encrypted.) Senators asked about user privacy and responsibility for content on Facebook, storing and selling of personal data, internet tracking of information, and potential election interference. As Vox put it at the time, "Senators seem to agree they want to fix something about Facebook. They just have no idea what."[3]

The hearing made it clear that there was not enough expertise in Congress in 2018 to understand the reach and impact of companies like Facebook. That's still true. Tech is complex, and most people don't understand in sufficient detail how "free" services make money (and in massive amounts) from users. A lot of lawmakers don't understand the specifics of this either and many other things about tech. Given that our expertise may be lacking, how are we to create impactful, public-focused legal and policy interventions that promote the public interest?

Too often, we rely on the private sector to help craft the laws and regulations that would limit them. This is especially true when there is insufficient expertise within the public sector to understand and project the societal impacts of complex technologies. Letting the private sector take the lead can create an inherent conflict of interest between profitability, a primary goal for business, and protection, a primary goal for the public sector. This conflict can perturb the realization and effectiveness of the law and policy under development. *The public sector needs to understand the technological challenges we face and not cede this expertise to the private sector.* We shouldn't be asking the foxes to guard the henhouse. Their job is to eat the chickens.

But as was demonstrated by the level of awareness at the Zuckerberg Senate hearing, technological expertise within the government is in short supply. For more than the last decade, the government has been trying to catch up. The United States did not have a chief technology officer until 2009, when Aneesh Chopra was appointed to the newly created position. The United States Digital Service (USDS) was launched in 2014 to supplement technological expertise in the government; it attracted technologists from the private sector interested in working with the federal government to apply their skills to public interest challenges. (Their early work is detailed compellingly in *ReCoding America*, which describes the challenges of implementing, and fixing, health.gov.)

At the beginning of 2025, the new Trump administration reorganized and renamed the USDS to be the Department of Government Efficiency (DOGE). DOGE's first order mission was to apply technical expertise to the investigation of government waste, fraud, and inefficiency, often through cuts to the public service workforce. By the end of 2025, thousands of tech experts had left the government, either through DOGE-related layoffs and firings or voluntarily. To recruit technology talent, the Trump administration created a new program in December 2025—Tech Force—to attract individuals with tech expertise to work in federal agencies.[4] The program seeks to attract engineers (especially those with AI skills) into the federal government for two-year stints. Tech Force is more similar to the early USDS than DOGE in its mission. Whether it is successful in attracting experts who can support the work of federal agencies and improve government technology systems is yet to be seen.

As digital technologies increasingly serve as critical infrastructure for the public sector, the need for tech expertise in the government continues to grow. It is widely known that the US government's digital infrastructure is often based on yesterday's technology and sorely in need of modernization. The Internal Revenue Service's Individual Master File, a system that holds all individuals' tax account data, was written in the 1960s in COBOL and Assembly language. Although modernization efforts have been underway for years, it is still being used midway through the 2020s.[5]

In all branches of government, technologists are needed to increase the government's in-house expertise. Technological expertise is needed to help develop policy and law in the legislative branch, modernize agency infrastructure in the executive branch, and stay on top of technological trends that impact global partnerships and the public across the government. An independent perspective is critical as the agendas of lobbyists and private sector representatives may conflict with the need for public interest protections. With the rise and risks of AI, increasing misinformation on social media, the growth of cryptocurrency, and consumer exploitation in cyberspace, there must be more technological expertise within the federal government.

Interestingly, we now actually have less capacity for providing independent expert technical advice in the legislative branch—Congress—than in the 1990s. In 1995, Congress defunded the Office of Technology Assessment (OTA), essentially closing the office. The OTA was created in 1974 to provide 1) a factual and nonpartisan foundation for policymaking in Congress, 2) a source of expertise to support Congress as it considered legislation relevant to technology and technological systems, and 3) early warnings of potential societal and other impacts of technologies and technological systems. The OTA shuttered its doors in its twenty-first year.

David Jensen was an analyst for the OTA in the early 1990s. His work involved policy research that evaluated and forecasted the impacts of technologies and technological systems, focusing on their short-term and long-term physical, biological, social, economic, and political effects—work that provided a foundation for assessing the usefulness of legislation in the public interest. Jensen and over 200 OTA colleagues worked full-time providing nonpartisan expertise to congressional committees, members, and their staffs.[6]

The OTA's staff created nonpartisan studies at the request of congressional committees on a wide range of topics. Their work allowed them to seek information from a broad set of stakeholders both inside and outside government, leveraging a treasure trove of deep expertise and information accessible to Congress. Jensen and his colleagues worked on studies about environmentally sensitive product design, the environmental costs of electricity production, and artificial intelligence technologies for detecting money laundering, among others.

For each requested study, the OTA's teams provided a thorough and nonpartisan analysis of the issues and their potential anticipated and unanticipated consequences. Interestingly, OTA teams provided no policy recommendations. They only provided findings about specific technologies being studied and a range of potential policy options consistent with those findings. It was up to the elected officials to debate and adopt specific policies. Once a report was developed or a study was done, the OTA team involved disbanded and each staff member moved on to other topics.

The OTA was small by government standards but important. Before its defunding in 1995, its budget was roughly $22 million (about .000015% of the approximately $1.52 trillion total of the 1995 US legislative authorization and less than one tenth the budget request of the Peace Corps that same year).[7] It was a bargain. Over the years, 140–220 staff, including Jensen, provided subject matter expertise in engineering, economics, law, physics, computer science, biology, and more. Over half of the staff had PhDs.

The year the OTA was defunded saw the launch of Amazon, eBay, Craigslist, Match.com, and the release of Windows 95 (Microsoft) and

Java (Sun Microsystems).[8] At a time when cyberspace was heating up, why did the OTA shut down?

The "Contract with America" was a legislative agenda introduced six weeks before the 1994 midterm congressional elections. It was signed by all but two Republican members of the House of Representatives and all of the party's nonincumbent Republican congressional candidates. Among its provisions was a promise to reduce congressional staff and make major cuts to the federal budget. Complying with this agenda put the OTA on the chopping block.[9] With a strong nonpartisan focus and weaker stakeholder support than other options, the OTA was defunded and has remained so for the last 30 years.

What has taken its place? In the wake of the closing of the OTA, technological expertise has increased elsewhere in the government. The Government Accountability Office's Science, Technology Assessment, and Analytics team (GAO STAA) focuses on technology assessment and analysis for Congress and technology readiness, in part filling the gap left by the OTA but without the structural independence, and with a different cultural and methodological focus. The National Academies National Research Council regularly performs studies at the request of Congress and executive branch agencies primarily with academics and some designated staff. Nonprofit think tanks (e.g., the Brookings Institution, Rand, the Heritage Foundation) perform studies on the implications of technology for policy. The Congressional Research Service, part of the Library of Congress, performs short-term analyses (in days, weeks, or months) and is primarily consultative—the typical OTA report was comprehensive and generally took around 18 months to produce. Tech industry lobbyists and academics regularly advise and testify before Congress. And federal agencies themselves have worked to increase their technical staff.[10] In the wake of the 2024 election, some staffing has decreased.

Can DOGE or potentially Tech Force supply the needed expertise once provided by the OTA? DOGE and Tech Force are part of the executive branch rather than the legislative branch. Moreover, DOGE's focus has been to identify waste, fraud, and abuse rather than the creation of studies relevant for tech-related legislation for Congress. The OTA had a different function than DOGE—expertise (still) needed by Congress to develop technology-oriented legislation effective for a changing world.

Is our current cache of in-house expertise available to Congress enough to guide the development of today's legislation, regulation, and policy governing the essential technologies of cyberspace? Many believe we need *more*: more independent, nonpartisan expertise, more full-time public servants, and more independent, thorough, public-focused analysis. There are a variety of options.

After 30 years, Congress could reestablish funding for the OTA with its existing charge or a revised charge to coordinate it with existing efforts, address its criticisms, and update its operation to deal with today's challenges. Alternatively, Congress could beef up existing sources of technology analysis and assessment from the Congressional Research Service, Government Accountability Office, federal executive branch agencies, advisory committees, the National Academies, or elsewhere. An in-depth analysis of the options by the Congressional Research Service indicates that the status quo is not enough—whatever choice is made, funding needs to be provided, and changes need to be made to provide the breadth, depth, independence, and thorough analysis offered by the OTA during its term of operation.[11]

Whatever happens will depend in part on politics and priorities in Congress, as well as how urgently other branches of government view the need for independent technological expertise. But what is clear is that *we cannot create good and effective laws without technology expertise*. To develop the societal infrastructure needed to reduce the most egregious risks of cyberspace, there must be expertise and knowledge infrastructure within the federal government focused on the public interest. Expertise that helps us understand the challenges in front of us, that helps create good laws, and that does not rely on the foxes to guard the henhouse for effective results.

CHAPTER 5

ENFORCEMENT: MAKING LAW, REGULATION, AND POLICY EFFECTIVE

Every parent knows that if you tell your kid not to do something and nothing happens when they do it, they will begin to ignore you—nothing will change. The same is true for laws. There are many examples of laws and regulations on the books that people ignore due to lack of, inconsistent, or weak enforcement. This is also true for laws and regulations in cyberspace—they must be effectively enforced to make a difference. If penalties or enforcement are insufficient, the law is less likely to foment change. Consider Vermont's Data Broker Law.

Data brokers are third-party companies that gather personal information, analyze it, and license it to be used by other companies for profit and often for marketing purposes. You can buy many kinds of datasets of individuals in specific marketing segments from data brokers, datasets that are often used to make decisions about you (and may contain inaccurate information). Many prepackaged datasets are shockingly focused on the vulnerable for targeting: "Rural and Barely Making It," "Credit Crunched: City Families," "Seeking Medical Care," "Rape Sufferers," and many others. The Rape Sufferers list was available in 2013 for just $.079 a name.[12]

In 2018, the same year that Zuckerberg testified before the Senate, the state of Vermont passed the first legislation targeting data brokers. The law required them to "register annually with the Secretary of State and provide information about their data collection activities, opt-out policies, purchaser credentialing practices and security breaches."[13] The goal of the Vermont Data Broker Law was to provide a public registry where Vermonters could find information about companies that sell their data and steps that they could take to delete their information.

But fines to data brokers for failing to register were relatively minimal—$50 a day and up to $10,000 a year—and enforcement was left to the attorney general's office with little additional infrastructure. Moreover, inclusion in the registry doesn't mean that data brokers can't continue to operate as usual; citizens must do the work, when possible, to remove themselves from data broker datasets. Then–Vermont assistant attorney general Ryan Kriger said that the state may go after companies that fail to register but described state efforts as "a toe in the water, not a cannonball."[14]

The Vermont Data Broker Law was an important first step toward more effective state and federal legislation to rein in the exploitation of individuals by the data broker industry. But stronger legislation, penalties, and enforcement are needed to address the problem—the sale and exploitation of personal information by data brokers who prey on vulnerable and other populations.

In contrast, the EU's General Data Protection Regulation (discussed in chapter 4) provides an example of how targeted legislation, substantial fines, and additional infrastructure can ratchet up effectiveness. For noncompliance, the GDPR sets forth fines for the most serious violators of up to €20 million, or up to 4 percent of the annual worldwide turnover of the preceding financial year, whichever is greater. To date, large fines have been levied on a spectrum of multinational companies, including the world's largest tech companies. In 2023, Meta (parent company of Facebook, WhatsApp, and Instagram) was fined €1.2 *billion* for transferring data from the EU to the United States without sufficient privacy protections. Two years earlier, Amazon was fined €746 million for targeted advertising not based on free consent.[15] In 2017, Google was fined €2.4 billion for giving preferential treatment in Google Search results to Google's price comparison shopping service over other services.[16] Other fines in the tens and hundreds of millions of euros have been leveled at H&M, British Airways, TikTok, LinkedIn, Clearview AI, and other companies for noncompliance with the GDPR.[17]

Coupled with the GDPR's fines is an infrastructure to oversee compliance and facilitate the adjudication of complaints. The legislation requires the identification of data controllers and data processors (individuals or organizations who make and implement decisions about the processing of personal data) and a data protection officer to ensure compliance with the GDPR for relevant organizations. It gives broad powers to independent data protection authorities to investigate complaints, intervene, and bring legal proceedings. The GDPR also creates a class-action-like option which allows data subjects to have their complaints brought by a non-profit dedicated to protecting their digital rights.[18]

In the United States, enforcement of existing privacy laws and antitrust laws does not generally involve this level of accountability infrastructure. Enforcement of privacy and antitrust violations against major tech companies by the Federal Trade Commission has varied by administration—depending on whether the current administration wants to take on privacy protections and/or the monopolistic market dominance of essential parts of cyberspace. In the US, these decisions are influenced by a more level playing field between the rights of private sector companies and the rights of citizens. This differs from the approach taken in the EU where there is a deep and constitutional commitment to support the digital rights of European citizens, even over the rights of corporations.

Enforcement is critical for effectiveness. Whether the enforcement mechanism is fines, restrictions on behavior, increased auditing and monitoring, expanded agency mandates, or some or all of the above, *something* needs to be done to ensure that the proactive interventions we use to rein in technology and reduce its risks actually work. Ideally, both due diligence in the form of sufficient expertise and analysis of societal outcomes and the incorporation (*and* implementation) of effective enforcement mechanisms to ensure compliance will result in successful law, regulation, and policy that drive positive change.

But success is contextual—successful law, regulation, and policy mean different things in different places. We consider the role of jurisdiction in creating successful interventions next.

ALIGNMENT WITH JURISDICTION: WHAT WORKS DEPENDS ON WHERE YOU LIVE

In the early days of the COVID-19 pandemic, China was remarkably successful in keeping COVID infection levels low and reducing the spread of the disease throughout the population. Its goal was zero transmission, an aspirational goal that we now believe is impossible, and its methods were strict.

Using China's pervasive surveillance infrastructure, citizens were regularly assessed as to whether they were likely to be COVID-free, at risk for COVID, or had COVID, using a green/yellow/red color code: "Green move freely; Red or Yellow, report immediately." They were asked to use an app that relied on personal identification information combined with location histories that showed their status. Citizens' apps were checked, and their temperatures were taken at a wide variety of health checkpoints, including subway and train stations, malls, and other public spaces. The color codes were used to permit or restrict citizens' movements: For those with COVID, quarantine was required, and pop-up medical facilities and food and supplies were readily available. People shifted jobs to provide staffing for the new COVID facilities and services.[19]

The World Health Organization praised the Chinese government's success in containing the disease.[20] But the success of the Chinese government depended in large part on its surveillance and citizen control infrastructure—a surveillance infrastructure that is used by the government to promote China's version of public well-being, an infrastructure that takes a vastly different approach to privacy than the United States and the EU. Would this approach have worked in more democratic jurisdictions? Not likely.

Not all countries view digital rights and privacy in the same way. Public sector interventions—law, regulation, and policy—vary by government ideology, jurisdiction, industry, and political power dynamics. Even within countries, digital rights and privacy may vary. California is one of the few states in the United States that guarantees a right to privacy in its state constitution. After the overturn of *Roe v. Wade* in the mid-2020s, the support for reproductive rights varied widely between US states, with vastly different rights afforded to women in Massachusetts versus those in Texas, for example.

For an illuminating contrast between jurisdictional approaches to digital rights, consider data privacy. The EU, United States, and China all have data privacy laws to promote the public interest, but what each jurisdiction considers to be in the public interest, what privacy protections are afforded, and what the government sees as its role in protecting the privacy of its citizens varies widely.

Start with the EU. In chapter 4, we discussed the General Data Protection Regulation, a landmark EU legislation that has influenced privacy laws across the globe. The EU's deep commitment to the digital rights of European citizens and the need for specific and effective protections drives the GDPR's comprehensive structure, detailed set of specifications, and high enforcement penalties. It also reflects the EU's governance structure—a supranational political and economic union based on representative democracy.

The European Parliament in general, and its GDPR legislation in particular, provides citizens and civil society organizations multiple ways to influence government decision-making. The ramp up to the GDPR and its preceding legislation was broadly consultative. This approach has continued to be applied to subsequent EU digital rights legislation, including the AI Act, and deeply reflects the interests and participation of European citizens.

You might be surprised to learn that China—a unitary communist state where surveillance infrastructure is an important part of its approach to governance—has privacy protections modeled in part on the GDPR. The Chinese Personal Information Protection Law (PIPL) went into effect in 2021. It incorporates its own version of many of the GDPR's protections in a way that aligns with China's governance and surveillance structure.

The PIPL focuses on people and organizations handling the personal data of Chinese citizens. GDPR-style themes are incorporated: The PIPL grants individuals the right to know about, decide on, or object to the use of their personal information except in specific cases like public health or national security. Personal data collection should be minimized and focused on a stated purpose. Individuals have the right to access, correct, delete, and limit the processing of their data. Some organizations are required to appoint data protection officers to promote compliance and there are large penalties for noncompliance.[21]

Although these themes are similar to the GDPR, they are implemented and embedded in a very different privacy infrastructure—one in which there is a deep and effective surveillance system that gives the government considerable access to the personal information of Chinese citizens. The purpose of China's surveillance system is to maintain social stability and national security and to promote a "harmonious society," all defined by the government.[22] Chinese companies are expected to cooperate with the government to support national security, public interest goals, and ensure regulatory oversight. Pay by credit card or chat with friends on the ubiquitous Alipay and WeChat services in China, and your data may be accessed and shared with the government to maintain state security and public order.

PIPL protections do not extend to the surveillance of Chinese citizens by government bodies, who are largely exempt when acting in the name of national security, public safety, and public health—areas that cover a lot of ground. They do provide privacy protections when dealing with foreign entities or private companies (except when working in partnership with the government). Although somewhat modeled after the GDPR, the close alignment between the PIPL's privacy protections and China's governance structure leaves state control and surveillance largely unconstrained.

In contrast to both the EU and China, the United States does not have an overarching GDPR-style national privacy law. Instead, the US—a constitutional federal republic with a strong emphasis on democratic representation and individual freedom—takes a sectoral approach with different laws governing the privacy of different types of data and different groups. Most US citizens are familiar with HIPAA (Health Insurance Portability and Accountability Act), which protects health data, FCRA (Fair Credit Reporting Act), which protects credit data, and FERPA (Family Educational Rights and Privacy Act), which protects student data. All are privacy laws that were created for specific sectors and specific needs.

The United States provides considerable latitude to its states to create a regulatory framework for their own residents. This has resulted in both legislation like the Vermont Data Broker Law and the California Consumer Privacy Act and its successor legislation, modeled on the GDPR.

In contrast to the EU's emphasis on individual protections, federal privacy legislation in the United States must align with a strong national desire to balance privacy with innovation and economic growth. Development of laws and regulations promoting consumer privacy often pit the interests of the American public against the interests and influence of large corporations. Efforts to promote (and enforce) federal privacy protections in the US must find a place in a system that is business-centric, sector-based, and fragmented. Although the US is moving closer to a more specific federal articulation of general consumer and citizen privacy rights and the laws that would protect them, we have yet to pass such legislation in both houses of Congress (or sometimes even move legislation out of committee) as of this writing.

To oversimplify (apologies . . .), the contrast between privacy protections in the EU, China, and the United States is a contrast between individual-centric, government-centric, and business-centric approaches respectively. And in the EU, China, and the US, as well as throughout the rest of the world, the more you know about how your jurisdiction's political and societal ecosystem works and what is important, the more likely you are to be successful within what is allowable to promote the public's interests and minimize the public's risks.

FOSTERING IMPACT

Expertise, enforcement mechanisms, and alignment with the prevailing political ecosystem provide a way to "prepare the soil" for successful law, regulation, and policy that promote effective societal outcomes. Good soil and a good environment are a great start for a productive garden. One more thing is needed: a good gardener. Someone who knows the terrain and knows how to promote, intervene, nurture, and make decisions that will cultivate the best outcomes. Who are the good gardeners for the public interest garden of cyberspace? Advocates and educators. We turn to them next.

6

HUMANS IN THE LOOP: ADVOCATES AND EDUCATORS

If Facebook is Goliath, Max Schrems might be David. In 2011, Schrems was an Austrian law student on a semester abroad in California when Facebook's privacy lawyer gave a guest lecture in one of his classes. He was shocked by the company's seeming disregard for Europe's data protection laws, a reaction that set him on a course that ultimately impacted his future, companies like Facebook, and two transatlantic data-sharing agreements between the United States and Europe.

Under the "right to access" provision of the Charter of Fundamental Rights of the European Union, citizens have the right to access and have rectified personal data concerning them (Article 8).[1] Schrems used this provision to request his data from Facebook and received a CD with more than 1,200 pages of the data he had volunteered on the site—likes, pokes, anyone he had friended or unfriended, past messages, and chats (including some he had deleted). In fact, Facebook retains even more data—data from third-party sources, public data, metadata (data characterizing the data), biometric data, and other information—but did not make that data available. Schrems was astounded by the amount of data he did receive: "The most prominent politicians had a thousand pages in their files. I'm just a normal guy who's been on Facebook for three years."[2]

Schrems redacted his data and posted it on a web page he called "Europe Versus Facebook."[3] The momentum built from there. He filed a series of complaints with the Irish Data Protection Commission (tax-advantaged Ireland is where Facebook is headquartered in the EU), accusing Facebook of improper handling of users' personal data. News articles on his advocacy and the europe-v.-facebook site encouraged 40,000 other users to also request their Facebook data. Facebook sent top execs to meet with Schrems in Vienna for a six-hour face-to-face meeting that left everyone unsatisfied.

Over the next decade, Schrems filed many more lawsuits targeting Facebook and other companies, including WhatsApp, Instagram, and Google.[4] Some of his suits were successful and some were thrown out of court. The suits were ultimately instrumental in taking down the "Safe Harbor" agreement between the United States and the EU in 2015 and later the "Privacy Shield" framework in 2020. Both were transatlantic data-sharing frameworks that required the US to protect the data of European citizens when transferred to the US as stringently as it would be protected in Europe. Arguments in both cases (sometimes called *Schrems I* and *Schrems II*) successfully asserted that these frameworks failed to conform to EU standards of personal data protection.[5]

The guest lecture that astonished Schrems in his early twenties set him on a course of advocacy that he remains on to this day. The GDPR gave Schrems and his colleagues more legislative tools to test how companies deal with the privacy of EU citizens' personal information and digital rights. Fines resulting from successful lawsuits include €1.2 billion levied against Meta for EU–US data transfers (2023), €50 million levied against Google for forced consent (2019), and €5.8 million levied against the dating app Grindr for illegally sharing user information with potential advertisers (2023).[6]

Today, Schrems heads up a nonprofit called NOYB—None of Your Business.[7] Launched in 2017, NOYB focuses on privacy rights, issues, and violations, using strategic litigation to force companies to comply with the GDPR and other EU digital rights and privacy laws. As of December

2025, NOYB had brought over 880 cases, 461 of which were pending, and was responsible for over €2 billion in fines based on NOYB complaints.[8] Some of the most recent lawsuits from NOYB take aim at AI. A series of 2024 suits against OpenAI, maker of ChatGPT, and Meta, parent company of Facebook, focus on how the personal data of EU citizens is handled in the training of AI systems and whether such training complies with digital rights protections in the GDPR. With the passage of the new EU AI Act, this portfolio of suits will no doubt increase.

Someone has to care to make things happen. Whether you think Schrems and NOYB are champions of privacy protections or bring suits that are "frivolous and vexatious," they have created momentum that has created change—change for Facebook, change for data-sharing agreements between the EU and the US, change that elevates the public interest in the public–private conversation.[9]

Advocates are catalysts for this kind of change. Max Schrems took on privacy protections. Ralph Nader took on vehicle safety in his book *Unsafe at Any Speed.* Arie Haagen-Smit took on air pollution in Los Angeles. Rachel Carson helped start the environmental movement with her book *Silent Spring.*[10] All started with passion, something to say, and an initial concrete step toward raising awareness and making change.

Although Schrems, Nader, Haagen-Smit, and Carson are well-known and highly effective advocates, the fact is that all of us can be advocates. You may have advocated for better food at your kid's school. I advocated for kayak storage at my condo. Some of us win big, some of us win small, many of us don't win at all. Whatever the outcome of a particular skirmish, advocates hang in there with the hopes of winning the war. Enough to put effort, energy, time, and personal resources into the issues they are most passionate about.

There are many advocates in cyberspace. They include advocates for companies (e.g., lobbyists) who would prefer that their organizations are less constrained by net neutrality policy, privacy legislation, requirements for AI transparency, and other interventions that seek to lower human risk but also limit the way companies operate, innovate, or make a profit.

Sometimes these advocates win: Laws and policies that promote the public interest don't get written, don't get out of committee, are used as a political football, or are watered down so they are ineffective. Sometimes these advocates don't win, especially when the public advocates for its own interests.

In today's world, the common good needs all the advocates it can get. That's where you and I come in. At the least, you and I can advocate for the things we believe in by the way we show up in the world: We can promote our interests in how we vote and who we vote for, what we buy and don't buy, whether we opt in or opt out, what we tell our children and our friends, what we promote at work, and so on. We can also advocate by creating proactive interventions designed to create change—the focus of Schrems' lawsuits, Nader's and Carson's books, my request to the condo board to approve kayak storage.

Both "insiders" and "outsiders" can proactively advocate for change. Schrems advocated for privacy protections outside of Facebook, using all the tools available to him within the EU's digital rights–focused body of legislation and policy. Many advocate from within their organizations to change how products are developed or how organizations operate. Jennifer Pahlka described many of these heroic figures in *Recoding America*, and she was an advocate for the public herself as part of the US Digital Service. Insiders, outsiders, and citizens making public-focused choices, *all* are needed to nudge cyberspace toward the common good.

You may be thinking: This is all well and good, but most of the things I care about are outside of my control. How can I advocate effectively? How could I make a dent in cyberspace? (Max Schrems might have been thinking the same thing during the Facebook lecture in his semester abroad class in California.) A number of years ago I read a book called *Switch* by Chip and Dan Heath that changed the way I thought about advocacy.[11] The book described a pragmatic and proactive approach that individuals can use to make the changes they want to see, even when they have little control. An approach that we can use to foment changes that can help nudge cyberspace more toward the common good.

FOMENTING CHANGE WHEN YOU DON'T HAVE MUCH CONTROL

If you want to advocate for change, you need three things: *You need to care about the change you want to make*, so much so that you are willing to keep going when you encounter the inevitable roadblocks or get discouraged or lose hope. *You need a plan*, a plan that incorporates a clear vision of what success looks like and the specifics of what you'll do to get there. And *you need to understand the characteristics of the landscape you're navigating* and how you might leverage its opportunities, address its challenges, and build community to foment change with you.

In *Switch*, the Heath brothers use the metaphor of an elephant, a rider, and their path to talk about these things. They talk about the importance of the *elephant*, or emotional component—your passion—in driving change. They pair the elephant with the *rider*—the part of you that executes a clear and actionable plan. And they talk about the *path*—how you might shape your ecosystem to achieve change. They point out that an effective approach to advocacy involves motivating the elephant to stay the course, directing the rider through specific and manageable actions, and shaping the path by creating an environment that facilitates change. Big changes require many efforts, some of which will be more successful than others, and some of which will feel like one step forward, two steps back. But successful social movements start from advocacy and often seem in retrospect like a "crooked line" to their goals. No worries. There are many paths to change. The whole point is to keep moving forward. (Indigo Girls fans will recognize the same sentiment from "Closer to Fine".)

Real change is a process and the result of many, often small, efforts: "Big changes can start with very small steps. Small changes tend to snowball . . . when change works, it's because the Rider, the Elephant and the Path are all aligned in support of the switch," observe the Heaths.[12] So how can you and I—public interest advocates in cyberspace—apply this approach? Consider some applications of the elephant/rider/path framework to the challenge of reducing the risks of the essential technologies in cyberspace.

Motivating the Elephant

When John Oliver focused on net neutrality in two episodes of his show *Last Week Tonight* in 2014 and 2017, he motivated so many citizen responses that they took down the Federal Communications Commission's website—twice. Oliver made net neutrality policy understandable and meaningful to a large community, and he generated genuine emotion about a policy proposal that included text like, "We propose enhancements to the transparency rule, adopting the text of the no-blocking rule from the Open Internet Order with a revised rationale, and creating a separate screen that requires broadband providers to adhere to an enforceable legal standard of commercially reasonable practices."[13] (Are you asleep yet?) How did that happen?

In *Switch*, three strategies are described that help motivate the public's passion and emotional connection with an issue ("motivating the elephant"): *finding the feeling* (nurturing the passion to create change), *shrinking the change* (making the change manageable), and *growing the people* (creating a community identity). Oliver did all three.

If you've seen *Last Week Tonight*, Oliver's Emmy-award-winning show, you know that each episode deals with a single issue and does an irreverent, fact-based deep dive. Oliver's first show on net neutrality in 2014 aired during the FCC's net neutrality public comment period and cheekily portrayed the relationship between near-monopolistic internet service providers and consumers as having "all the ingredients of a mob shakedown."[14] After the show, comments on the FCC's public site ballooned to over 47,000 and contributed to "technical difficulties" in the FCC servers hosting the site (i.e., they crashed it).[15] The FCC eventually ruled in favor of net neutrality, in part because of administration support and the strong public response.

Oliver's second show on net neutrality aired in 2017 after a change in presidential administrations.[16] The FCC again considered net neutrality, this time with an eye toward rolling back the policy adopted in 2015. (The FCC chair at the time, Ajit Pai, claimed in the press that net neutrality's "days [were] numbered" and vowed to take a "weed whacker" to the policy.[17]) Oliver revisited the subject in 2017 during the public comment period and created a website—gofccyourself.com—that simplified the complex process of leaving a comment on the FCC's official site. Within minutes of the broadcast, traffic to the FCC website spiked by 3,116 percent "resulting in the disruption of system availability."[18] The FCC initially claimed that its servers went down due to a denial-of-service attack, but an investigation showed that the spikes in web traffic coincided exactly with the timings of Oliver's broadcast on various media.[19] Although the comments were strongly in favor of retaining net neutrality, the Trump administration did not support it and the Obama administration's net neutrality policy was repealed . . . until the next change in administration.

To create change, you win some and you lose some, but you can't do either without caring about the issue. Oliver "motivated the elephant" by helping viewers understand how net neutrality would impact them and why they should care. He then made it easy for them to respond to the FCC's call for public comments—he shrunk the change by simplifying the process of making a public comment to the FCC via the gofccyourself website. As part of his deep dive, he created a community identity—he grew the people who cared about net neutrality, and he encouraged them and gave them tools to express their views. He helped empower the public to advocate for their interests.

Both episodes spurred public engagement on net neutrality policy. Politics still matters a lot—the presidential administration has a lot to do with which policies are promoted—but substantial public feedback is rare. Oliver succeeded in galvanizing a more substantive discussion of the risks and impacts of an essential technology—the internet—and motivated a broad community to weigh in on net neutrality.

Directing the Rider

Change doesn't just happen; you need a plan. In chapter 2, we saw how a coordinated and committed set of citizens and groups—#Oakland Undivided—did the nearly impossible during the COVID-19 pandemic:

They dramatically narrowed—almost closed—the digital divide when remote learning was the only way to attend school. When only one in four of the district's elementary, middle, and high school students had adequate devices or connectivity for remote learning, #OaklandUndivided developed a highly specific and coordinated set of small, manageable actions to close the gap between the district's digital haves and have-nots. In short, they "directed the rider" to create results.

Directing the rider involves three important strategies: *following the bright spots* (investigating what's working and cloning it), *scripting the critical moves* (determining specific actions and behaviors that are needed), and *pointing to the destination* (having a clear vision of what success looks like). The #OaklandUndivided consortium did all three.

The clear vision for the consortium was to close the digital divide for Oakland students, and as quickly as possible. Success was reaching every school-eligible child in the district and giving them the tools they needed to remotely attend school. Although more extensive broadband access would help permanently address the problem, there wasn't time to make this happen quickly. #OaklandUndivided mobilized district, public, private, nonprofit, and citizen groups to work together in a highly scripted and choreographed way to distribute over 36,000 computers, create 11,500 hotspots, and engage with families in the district. They replicated effective strategies again and again. They got astoundingly close to completely closing the gap. #OaklandUndivided's effective planning, driven by passion and advocacy, and their highly coordinated efforts achieved results and created momentum.[20]

Today, Oakland continues to direct the rider, persisting in its efforts to increase Wi-Fi and broadband in the city and close the digital divide. A Broadband Master Plan, approved by the city council in May 2025, guides the development of "Oakland Connect," a city-led project to build a municipal broadband network using a hybrid public-private model. The plan calls for the network to be operational by 2027, moving Oakland closer to the goal of providing fast, affordable internet access to all residents.[21]

Shaping the Path

Sometimes change is hard because you're swimming upstream. Having the stream carry you forward as opposed to fighting the current can make a big difference for creating and maintaining positive change. Shaping the path means you can facilitate change by making it easier and more habitual. Three strategies can shape the path to make positive change easier: *tweaking the environment* (making the environment support the change you want to see), *building habits* (operationalizing the change through specific behaviors), and *rallying the herd* (growing the community of people supporting the change).[22]

Weight Watchers, Alcoholics Anonymous, and other groups are masters at shaping the path. For everyone who has ever been on a diet to lose weight, we know how hard it can be. Stopping eating is not a sustainable solution. Forgoing the foods you love is hard. Most diets are ultimately abandoned and don't lead to lasting results.

Today, new weight loss drugs are redefining how individuals approach weight management. But for many years before, Weight Watchers employed a successful behavioral method to shape the path to weight loss. By creating a point system that assigns low points to healthier foods (fruits and vegetables for example, low points mean you can eat more of them) and higher points to more fattening foods (candy bars, ice cream), participants become more accountable for what they eat. Weight Watchers builds a support system (rallies the herd) of others doing the same thing who help you stick with it. Tweaking the environment, building good habits, and finding a community to do it with promotes success. Alcoholics Anonymous uses similar techniques, tweaking the environment and building habits through a 12-step program, and building critical individual and community support for abstinence. For participants in these programs, everywhere you look, the path has been shaped to help you succeed.

We shape the path in cyberspace when we make the right thing to do the default, easy, habitual. Our systems can help us. Two-factor authentication is a default that makes your information more cyber-secure. Using the built-in Time Machine backup feature on my Mac with an external hard drive is easy and helps me preserve my valuable data. My machine proactively tells me when there are critical security patches and important updates for the system. When I get the alert, it's become a habit to update my computer.

Although shaping the path involves extra effort on the part of companies in terms of alerts and features, and on the part of users to proactively do the right thing, it works. Providers and users could do even more. Companies regularly shape the path for commercial products to maximize profit. They could shape the path for digital critical infrastructure to better promote low-risk user behavior. For example, greater transparency about where information comes from and algorithmic priorities, along with stronger privacy and security protections, could help users engage more responsibly with the digital technologies they depend on.

Max Schrems used all of the advocacy strategies described here: He motivated the elephant, directed the rider, and shaped the path to promote personal privacy protections. He used his passion (the elephant) to grow a community of fellow travelers through public advocacy and the europe-v.-facebook website to join him on his quest. He used strategic lawsuits (directed the rider) toward a clear goal—making Facebook and other companies comply with EU data protection laws. His portfolio of lawsuits (scripted moves) increased the pressure on Facebook and other companies to comply with EU privacy laws. He shaped the path by strategically leveraging EU data protection authorities in multiple EU countries, the GDPR's toolkit of provisions, and a growing community of concerned citizens to promote privacy rights and accountability. His advocacy fomented change.

The gap between the societal outcomes we see and the societal outcomes we want to see can drive public advocacy in cyberspace. Your understanding of how tech impacts the world and your awareness of the technological and societal environment in which you would like to see change happen can drive effective courses of action. Experience with planning your path and visualizing your desired results can help you move forward more effectively. You can do all of this with little or no preparation (from scratch), but you will get further faster if you start with some awareness of and experience with the sociotechnical world in which you must operate.

Where should you gain this awareness and experience? In school. Today's education should provide students with the knowledge and tools they need to successfully solve problems and navigate the world they live in. A world that is tech-enabled and where social responsibility is critical for society to thrive. To help the advocates of the present and the future workforce effectively lower the risks of cyberspace, we must begin to teach students about the responsible creation, use, and management of technology as a core competency in our schools.

EDUCATING SOCIAL RESPONSIBILITY FOR A TECH-DRIVEN WORLD

Today's students have lived in cyberspace since birth. Their baby pictures have flooded Facebook and Instagram. Many got their first smartphones before adolescence. They are more likely to scroll on social media than to pick up a book. They are both jaded about technology and increasingly worried about where cyberspace and the organizations that contribute to it are going. They want their jobs and their lives to promote their interests. In a 2024 study on millennials and Gen Z, Deloitte identified an important trend: "Three-quarters of Gen Zs and millennials (75%) say that an organization's community engagement and societal impact is an important factor when considering a potential employer." Moreover, "the majority of Gen Zs (86%) and millennials (89%) say having a sense of purpose is important to their overall job satisfaction and well-being."[23] It's not just about money.

A lot of today's college students are looking to make a difference in the world. They want their schools to train them for a more socially responsible future. Students are concerned about today's often overwhelming problems: political polarization, global warming and climate change, the rise of extremism and violence, racism and exploitation, the encroachment of AI on the workforce. These students are inheriting a world they didn't create, a world in which people often come last—after politics and profit. They want to do something to change that world. Truth be told, so do I.

The challenge for educators has always been to prepare students to succeed in the world in which they live and address the challenges of the day. Today's curricula must prepare our students to understand and keep up with the rapid technological advancements that are only going to become more prevalent. It is important that students build their own skills and knowledge; it is also important that they be able to capitalize on tech's opportunities, assess its risks, and develop effective solutions to address its challenges. We must prepare them to become knowledgeable professionals and citizens and to manage technology for benefit in their work, their communities, and their lives. This means that what we offer as educators must change.

How can we prepare these students and what do they need? We can help them develop core competencies for the world they live in: *tech literacy* (a working knowledge of what tech does and how it operates), *social governance strategies* (interventions to guide the beneficial use of tech), and a *public interest tech mindset* (a way to think about technology that promotes public well-being and reduces risk). With these core competencies, students will have a foundation for successfully navigating today's sociotechnical world and the current and future opportunities and challenges they will encounter.

Take tech literacy. Most students are not computer science majors, but it is important for *all* students to have a working knowledge of technology design, function, and impact on their work and lives. Tech-literate students don't need to know how the operating system of their smartphone works. They do need to know that their smartphone and its apps collect personal data, use this data to make decisions, and may share, exchange, or offer up this data to third parties based on company policy and prevailing legislation. Tech literacy is fundamental for the world we live in to effectively assess the outcomes and risks of the technologies we use.

Students need similar knowledge about social governance strategies and interventions. As we've seen throughout part II, law, regulation, and policy, as well as socially responsible innovation, design, and development can all be used to drive tech that advances public well-being and minimizes human risk. Knowledge of what tech can do (tech literacy) and what people can do (social governance strategies) is fundamental as we try to guide the essential technologies of cyberspace to put the public first.

Finally, it is important that students develop a public interest tech mindset, as well as a sense of what public interests are (e.g., ethics, sustainability, privacy, safety) and how they can be promoted alongside other interests. The responsible use of tech is rarely about good guys and bad guys. It is about the myriad of choices we face in developing, managing, and using tech, and their anticipated and unanticipated outcomes. Experience with making these choices, especially in real-world environments in which other priorities are important, is critical.

Practice with challenges in real-world settings (sometimes called *experiential learning*) is important for students. In our complex societal ecosystem, decisions are messy. The meritorious/best option doesn't always prevail. Sometimes it's more profitable, someone else's turn, or more convenient or politically astute to make suboptimal choices from a public interest perspective. Students need experience with this common scenario; they need to learn how to use and develop strategies that promote both business goals and public protections, especially when they conflict. Universities can provide this experience through internships, project work, service, and other means. When students graduate with this experience, they are more likely to be successful in using technology responsibly in their careers, whatever those careers may be. Our era demands this. As Deloitte indicates, students want this. All signposts point in the same direction.

Over the last decade, there has been a groundswell of interest in incorporating tech and social responsibility into higher education. The think tank, New America, working with the Ford Foundation and others, recognized and capitalized on this interest by creating the area of public interest technology as "the study and application of technology expertise to advance the public interest, generate public benefits, and promote the public good."[24] They launched the first public interest tech consortium of universities—the Public Interest Technology University Network (PIT-UN) in 2019 to bring together higher education institutions, educators, and students to build public interest technology as a specialization and "grow a new generation of civic-minded technologists."[25]

My own university, UMass Amherst, joined PIT-UN in 2021–2022. Within the almost 50 universities in the consortium at the time, we found fellow travelers and many approaches to incorporating public interest tech into curricula and practice. On our campus, we created the Public Interest Technology initiative (PIT@UMass[26]) and took on the challenge of conceptualizing what a *general education* approach to public interest technology instruction might look like: What if public interest technology were part of every major and taught with other core competencies? How could we help students apply a public interest tech mindset and skills to nursing, finance, humanities, computer science, public policy, engineering, and other areas of academic pursuit? We created a general education course—Introduction to Public Interest Technology—open to all undergraduate students and majors on campus (a syllabus is included in the appendix). We created PIT certificates to document the building of responsible tech skills and awareness that undergrads, grad students, and remote learners could pursue.

Other programs targeted faculty and the broader community. We worked with faculty across campus to promote research that incorporates responsible tech strategies. We partnered with local communities to build AI literacy and employ AI safety practices. The goals of PIT@UMass's coordinated portfolio of programs? 1) Build knowledge, skills, and experiences that positively shape the way individuals and communities engage with technology, and 2) train a workforce with the skills to use technology responsibly and the expertise, strategies, and commitment to put those skills into practice. As our programs continue, the PIT@UMass initiative grows in reach and impact.

Pragmatically, creating public interest tech programming is not easy. Because there are many different approaches to promoting responsible tech, public interest tech often suffers in an academic environment by being too broad. Moreover, for faculty, the impact measures of public interest tech are often not well-aligned with scholarly measures of impact: Public interest tech is more about practice than theory, more focused on action and service than models and observation. As an emerging specialization, it's short on the academic infrastructure—grants, conferences, publication venues, writers of tenure letters—needed to satisfy traditional promotion criteria for faculty to achieve mainstream academic success.

An additional problem for public interest tech programs is funding. Grant funding typically targets innovation rather than the infrastructure needed to run programs. Because of this, sustainable PIT programs generally require institutional funding to achieve impact. Although it's hard to find people on a university campus (or just about anywhere) who think social responsibility is a bad idea, universities have many competing priorities for funding. Faced with constrained budgets and many mouths to feed, academic institutions are more likely to invest in innovation than social responsibility, much like other organizations.

But to succeed in their missions, universities must evolve to meet the challenges of the times and prepare students for the present and future sociotechnical world in which they live. Despite the pragmatic challenges of funding and aligning academic and public interest technology metrics of success, more academic institutions are making room for "tech and society," "ethical tech," and "responsible AI" courses and other public interest tech–relevant courses and programs. The academic tanker is beginning to turn.

For tech to be better and better for us, public interest tech programs are needed to train today's and tomorrow's responsible tech citizens and workforce. Programs like the PIT@UMass initiative and those of other institutions are starting to equip future citizens, professionals, and leaders with the skills they need to promote human well-being in today's tech-driven world. More such programs are needed.

You don't have to pass a law to promote the common good. You *do* have to make public-focused choices. There are many ways to make tech better and nudge cyberspace away from unacceptable risk and closer to the common good, from socially responsible innovation and people-centered development, to effective public interest legislation and policy, to advocacy and education. All of us can be involved in something that makes a difference—small or large. There are many options.

If more of us adopt a public interest tech mindset, how could that impact the workforce? How could that reframe a world in which AI will increasingly call the shots? We explore those questions in part III and the final chapters of this book.

III

PUTTING THE PUBLIC FIRST

PROLOGUE, PART III

When President Trump took office in 2025, the American political and societal landscape changed dramatically. National priorities and new tariffs shifted the geopolitics driving tech innovation and leadership. The new administration's cuts, priorities, and increased proactive control of research dollars directly challenged the business model and mission of US universities. The landscape of opportunity changed.

As I write this, the responsible creation, use, and management of tech for the common good—rarely a dominant or high-profile priority—seems more elusive than ever. Public interests—fairness, security, safety, privacy, sustainability—are being viewed through a different lens, effectively swapping two broad and polarized groups of Americans: those who feel that the right kinds of public protections are now under attack, and those who feel that the right kinds of public protections are now being put into place.

However public rhetoric and priorities change, cyberspace still constitutes the ubiquitous ecosystem of the world we live in, and its digital critical infrastructure is still rife with risk. Ignoring these risks, talking ourselves out of them, or making them look less consequential will not make them go away. Our actions matter, now and in the future. If we ignore today's warnings, we will nonetheless suffer their consequences despite assurances that we are on the right path. We must move beyond rhetoric to deeply assess the risks of the technologies on which we depend, and we must act to benefit our communities, our children, our world, and our future. As always.

The change in US presidential administrations reminds us that societal interventions—guardrails, policies, laws and regulations, priorities—can be done and undone. Perspectives can change. Realistically, we have always engaged with technology in a dynamically changing societal and political environment, one that makes some things easier and some things harder. We must work in the context of the time we are in, but we shouldn't take our eyes off the ball.

In particular, *the public interest–focused strategies described in this book are more important than ever.* They provide tools to manage our present and future and to promote human well-being in a sociotechnical world. Our efforts—small, medium, and large, in all sectors, by all stakeholders—are essential to create and ensure that we have digital critical infrastructure that empowers rather than endangers us. The cyberspace we experience and create should advance personal freedom and public protection, not science fiction: We don't want unfeeling robots to take over the world, nor do we want to live in an environment that feels like *The Minority Report.* Now is the time to secure the future we deserve.

There are many differences between 2024 and 2025 and beyond, but cyberspace is still the ecosystem in which society functions and we must use our agency and humanity to guide it toward societal benefit. The strategies in part II provide examples of actions spurred by a public interest tech mindset, strategies that are now more relevant than ever. Progress won't be linear, but really, it never has been.

Part III focuses on where we are at this moment. We consider how we might parlay the present into a future where tech is better and better for us. In chapter 7, we consider public interest tech "in the wild"—the responsible use of tech for today's work and in today's workforce. How can we meet the societal challenges of tech in the work we do and the work we are beginning to rely on AI to do?

In chapter 8, we explore how we can prepare for governance in a hybrid society, one where decisions are made by both people and autonomous "intelligent" machines and where legal notions of responsibility and accountability must apply to the consequences of both. How should we evolve governance to ensure that humans will thrive in a hybrid world?

Finally, in chapter 9, the final chapter, we step back and reconsider the big picture: How can we reframe the sociotechnical world in which we live to be better? How should we think about its ethnology—the relationships between its human and automated actors? How should we think about its ecology—its digital critical infrastructure and invasive technologies? The choices will be up to us, as will the benefits and risks of our actions.

7

PUBLIC INTEREST TECHNOLOGY IN THE WILD: PIT AT WORK AND IN THE WORKFORCE

Scan the job boards and you will find few jobs with the title "public interest technologist." No matter, a public interest technology mindset and strategies can be used in any job. In the same way that you are less likely to have a job as an "ethicist" but can apply a sense of ethics to any job, public interest technologists employ responsible tech strategies in whatever job they do, regardless of title. In this way, a public interest technology mindset serves as more of an "engine" (powering from the inside) than a "chassis" (defining from the outside) for modern professional work.

In part II of this book, we described many ways our workforce and our work could be "PIT-focused" by strategies that help us deal with digital technologies more responsibly. We can use a public interest tech mindset to employ responsible approaches to the design, development, operation, and use of technology (chapter 3); develop societal controls such as law, regulation, and policy to lower human risk (chapter 4); promote enforcement strategies that ensure that laws and policy are effective (chapter 5); and advocate for responsible tech in our jobs, communities, and lives, as well as build public interest tech awareness and skills into education and training programs (chapter 6).

A public interest tech mindset is even more important today when our work is increasingly influenced by big global and social forces: *climate change* (impacting how we do our work and how tech is sourced and powered), *long-term impacts of the COVID-19 pandemic* (affecting the prevalence and use of tech for remote work), a *generational workforce transition* between retiring baby boomers and millennial and Gen Z workers (impacting what workers expect, what they know, and what they must learn), and *regime and political changes* (influencing the societal and geopolitical environment for trade, collaboration, and competition). These forces are coupled with major technological forces: *AI's increasing encroachment* on the workforce, the prevalence of *data collection and surveillance* in work and life, and the societal and environmental impacts of using and managing *tech at scale.*

All of these forces exacerbate the need for today's and tomorrow's workforce to *work differently*—to make, manage, and use our driving technologies more responsibly, to realize tech benefits while keeping humans in control, to ensure that the work we do advances the society it supports.

In this chapter, we explore public interest tech "in the wild," the environment where the strategies from part II play out—the real world of work and life. We will meet Kathy Pham, one of the most inspiring public interest technologists I know, and consider three major tech-driven trends impacting today's work—surveillance, scale, and smarts. We lead (as always) with the humans in the mix.

RESPONSIBLE TECH WORKERS: A PUBLIC INTEREST TECHNOLOGIST IN THE WILD

Today's public interest technologists work in all sectors and in a wide variety of jobs. Their impact depends on how successfully they are able to suffuse public interests and responsible tech strategies into their organizations and work. They are passionate about making things better. What drives them? Meet Kathy Pham.

Pham is a public interest technologist, even though that hasn't been the title of any of her jobs. She's hip, accessible, wicked smart, and has been on the frontlines of tech for most of her career. She currently works in the private sector as vice president for artificial intelligence at Workday, but her previous jobs have spanned government, the private sector, nonprofits, venture capital, and academia. Wherever she's worked, she's made a positive difference.

What makes Pham an inspiring public interest technologist is her *approach*, the way she looks at problems holistically and elevates the public interest in their solutions, and her *impact*, the success she's had in moving organizations and their efforts toward greater social responsibility and public benefits.

Pham's parents were refugees who were sponsored to come to the United States after the Vietnam War. Born in the US, she grew up knowing the power of government policies to impact the trajectory of lives. By the time she was 10, Pham was helping her community fill out job forms, manage immigration processes, set up bill payments, and communicate with social service offices—often as part of the transition from one home and familiar ways of doing things to a new home and new ways of navigating in a new country.

Technology was becoming more pervasive as Pham was growing up. She wondered whether tech could be more helpful in getting people what they needed, whether tech could bring more dignity to the social service environment. She developed a passion for making technology work for people—to make it easy to use, more empowering, and a positive experience for those who engaged with it. Pham majored in computer science at Georgia Tech and created a public health startup even before she graduated. She received offers from big tech companies—Google and IBM—an aspiration of many computer science students at the time.

During the early 2010s, Pham was drawn to the US Digital Service and the opportunity to help people like those she grew up with access services and successfully navigate the US landscape. It was difficult work but enormously satisfying. She sometimes found the societal and organizational problems of getting things done just as important as the technical problems, echoing the experience of many technology professionals in government. When Pham left government, she brought her experience and skills to Harvard where she teaches a course in product management and society, helping students build technology responsibly. As VP of AI at Workday, she promotes the public interest in her technical focus, her priorities, how she manages her team, and how she thinks about her customers.

Pham has put into practice all the strategies in part II—innovating responsibly, shaping responsible development and design, advocating for responsible tech, and working in the public and private sectors to make tech responsibility a priority. She is a bright star in the public interest tech workforce. There are many others—in companies, nonprofits, education, government, and elsewhere—who successfully elevate the prioritization of people and the responsible use of tech in their jobs. They seek to change the way we operate in order to lower tech risks and promote better tech outcomes. That's part of my job as an educator. Whatever your job, that can be part of your job too.

The work that Pham, you, I, and other public interest technologists do is driven by the speed of tech. Over the last several decades, information technology has transformed work in every industry and sector. Innovations in computer "feeds and speeds," collection and use of data, the extraordinary growth of AI, and many other advances have supercharged businesses, organizations, and sectors, and are driving the dynamics of work and our way of life. It is within this environment that public interest technologists must operate and succeed. There is no specific recipe for doing so, just like there is no specific recipe for being an ethical person. Public interest technologists employ a public-focused mindset in their work, as Pham does, and use their skills, awareness, experiences, and multiple strategies to promote human well-being.

How do we apply a public interest tech mindset to today's work? We explore this question in the remainder of this chapter with a focus on three key trends shaping today's work and workforce: *surveillance, scale,* and *smarts.*

Each of these trends has a tremendous impact on modern work and life: Surveillance and the ability to collect and analyze massive amounts of data have changed the nature (and human risks) of the work we do. The growth and scaling of today's successful businesses and the size of their customer bases have fundamentally expanded their power in the market, influence on governments, and modes of engagement with the public. Advances in the capabilities and applications of AI have ushered in a new era where our work increasingly has "AI inside" and sometimes "AI instead of." Unarguably, these trends are transforming the work we do as well as who is doing it.

In the next sections, we focus on the challenges and opportunities of responsibly dealing with surveillance, scale, and smarts for tech systems, services, and products. We begin with the growing prevalence of surveillance and its potential to transform our work and societal environment from one in which we are relatively free to operate in private to one in which each of us is increasingly being watched.

INCREASING SURVEILLANCE: IS IT USEFUL OR INTRUSIVE?

The subtitle of Shoshana Zuboff's 2019 book *The Age of Surveillance Capitalism* is *The Fight for a Human Future at the New Frontier of Power.*[1] The book makes a detailed and compelling argument that large-scale collection of data turns surveillance into an effective tool for capitalist accumulation, manipulation, and exploitation. This is not news in 2026. More than half a decade after the publication of Zuboff's iconic book, data collection via clicks, text, sensors, cameras, biometrics, and the like is even more prevalent in our work and life. For the most part, we have just . . . gotten used to it.

Sourced from the data we provide directly (through clicks and information we volunteer as we navigate the internet) and indirectly (through the accumulation of metadata and the purchase, sharing, and exchange of other data related to us), surveillance data on our whereabouts and activities is increasingly part of our massive data profiles. Drive west on Interstate 90 from Boston, Massachusetts, to Albany, New York, and the license plate of your car (and possibly the images of drivers and passengers) is captured in both Massachusetts and New York by automated license plate readers at regular intervals. The RFID tag on your E-ZPass communicates with the readers to make it easy to pay the toll. If you're speeding, the image of your car captured by the cameras does not *yet* result in a ticket, but it could. It has all the information law enforcement needs: Your car somehow went from point A to point B in less time than the speed limit permits.

Go to the mall and cameras capture your image. Facial recognition can mostly determine exactly who you are (with some bias and high-profile errors). That's also true in buildings, parking lots, and cities around the world. What is this data used for? Is this capturing of your location protective or intrusive? It depends. Yes and no to both.

The digital surveillance industry was valued at $130 billion in 2022 and is expected to grow over 80 percent—almost doubling—by 2027.[2] In addition to using surveillance for safety, threat detection, and monitoring of license plates or who is at the mall, surveillance systems are increasingly deployed *by us*. Today's smart doorbells, in-home cameras to monitor pets, infants, or elderly relatives, and systems used to show real-time information, such as the surfcams in San Diego or construction webcams at new building sites, are all examples of voluntary adoption of surveillance.

Many consider this voluntary surveillance to be protective and informative, and it often is. But sometimes the collection of this information can be intrusive or worse. A student told me that she was concerned about being captured on a real-time campus construction webcam that included images of a parking lot because she was being stalked in her private life. She was concerned that the stalker would be able to see when she was on campus, gaining information about her location she did not want him to have. Although surveillance technologies are often used for protection or useful information (Is the surf good today?), nearly all of these uses of passive surveillance do not require the approval of those in their view—something that is likely infeasible. And there is often little transparency as to when and how these surveillance systems are being used, the longevity of their data retention, who has access to the data, and what the scope of their capture is.

Surveillance from cameras and other devices coupled with facial recognition and data analysis doesn't just allow us to track that someone is there; it can increasingly tell us who that individual is. This potent combination is beginning to change how we do our work. Consider law enforcement. The use of surveillance, facial recognition, and data analysis are powerful tools that have led to the charging of both criminals and innocents. In chapter 2, we noted three false arrests based on biased facial recognition systems and insufficient scrutiny by a "human in the loop." To mitigate risk, the use of surveillance must be coupled with human judgment and adequate police work to eliminate potential suspects—the basic investigative work by law enforcement that includes checking alibis, comparing tattoos, following DNA and fingerprint evidence at the scene, and other techniques before making an arrest.[3] Human judgment is also imperfect, but surveillance must add to both our tools *and* our accuracy for full benefit. Using a public interest tech mindset means that the use of surveillance tools must be overseen by careful humans who can vet and assess their results to ensure the protection of their subjects.

Sometimes surveillance is the wrong tool for the desired outcome. During the COVID-19 pandemic, many companies moved to remote work to accommodate the closing of on-site offices and workers who chose to move to different locales—nearer to family, with more affordable housing, with more yard space for the kids, and other benefits. As COVID moved from pandemic to endemic, remote work became institutionalized in many companies, with over 20 percent of employees working solely remotely by 2025 and over 50 percent of workers working remotely at least part of the time.[4]

Supervising remote workers is hard, both in terms of ensuring adequate security for the work and in terms of monitoring the workforce. During the pandemic, many companies required remote employees to use VPNs (virtual private networks) and specific privacy and security software, or specifically outfitted work computers to ensure the protection of sensitive content. Many also employed *worker surveillance* tools—systems installed on your laptop that monitor computer usage for remote employees. Such tools often collect massive amounts of data—work *and* nonwork—during remote work periods: all the windows you open, your browsing history, screenshots that may include texts you respond to, always-on cameras tracking your attention, and other ways to record your behavior. Such monitoring was, and is, used to ensure that employees are working when and on what they are supposed to, not just playing Minecraft. For some, monitored information could also be used to feed performance assessments and worker scoring systems. Many employees put up with surveillance because they want the job, but their work is negatively affected by what they consider "spy tech."[5]

Is this intrusive? Is it ethical? In 2024, the US Government Accountability Office remarked in its blog that "digital surveillance tools are putting employees on edge and having other negative effects that could impact their mental health, morale, productivity, and more." It pointed out that surveillance technologies could cause additional stress and magnify disability and other disparities in the workplace.[6]

In the case of worker surveillance, using a public interest technology mindset means that the organizational leaders seeking a productive workforce must also prioritize the privacy of their employees. They must consider more broadly other impacts of the use of worker surveillance technologies, including the stress levels and psychological well-being of employees. Taking a more responsible tech approach to oversight may mean that employers use less invasive tools for performance monitoring and in all cases ensure that employee protections are in place and privacy is respected.

INCREASING SCALE: DEALING WITH MORE THAN HUMANS CAN HANDLE

As the world population gets larger and digital goods and services become more prevalent and multinational, the customer bases for the most popular and essential digital technologies have become huge—sometimes larger than the size of any individual country. In 2025, India had the largest population in the world—close to 1.5 billion people. In that same year, Facebook had twice that—over 3 billion users; Amazon had over 300 million users, more than the 2025 population of Brazil. Businesses with large amounts of information and large consumer bases must find efficient ways to manage operations and engage with their customers. They turn to digital technologies when it is not feasible or cost-effective for humans to do everything.

Digital technologies can introduce great efficiencies at scale, providing a way to streamline operations, production, service delivery, and workflows. They can reduce human error, manage extraordinary databases of information, and both market to and support customers. Automation and AI can help triage, directing some decisions to systems that can deal with common problems and some to humans who can deal with the most challenging issues more effectively.

But sometimes automated systems necessitated by scale don't do a good job providing what customers need. The public interest challenge in dealing with scale is to know when tech-driven efficiency is winning over effectiveness, to ascertain when customers are not well-served by automation, and to do something to more effectively prioritize people in this scenario.

For me, there is no better example of this challenge than today's scale-driven automated customer service systems. It often feels like some of these systems are actually best at separating the humans in the database from the humans who can help with their problems. I ask the reader to indulge me for a bit with a personal rant about some of my own worst examples of automated customer service.

<rant> Customer Service vs. Custom Service: Are We Really Better Off with Automation?

- I am on hold with music that is no doubt the same music played on loop in the fifth circle of hell. It's been 35 minutes. It will be 50 minutes before the system will hang up on me because I am not a leaf in its automated customer service decision tree. I am angry and frustrated. There is no way to talk to an actual human.

- My credit card company referred me to a monitoring service when its data systems were breached (again). The monitoring service sends me a sexual predator alert about my file. What?? I can't get into my account because my password won't work and I have to be on the site to change my password.

  I call the customer service number and try to get myself routed to a human by saying the word "agent" 25 times. By the time I get to an actual human, she sympathizes and says that the magic word is "fraud"—you always get to a human immediately because in the case of fraud, the credit card company is the victim. The fraud humans can route you to other humans who can help me change my password. It turns out that no one thinks I'm a sexual predator, I just live in the vicinity of one.

Sound familiar? In the increasingly automated world of customer service, *custom service*—actual humans who can solve problems and answer questions—is on the wane. What we increasingly get are automated customer service systems that are frustrating to deal with and don't solve our problems. Most people call customer service only when something is not working, but it seems that these systems are designed to address "common" problems—problems that you don't necessarily have. When there are no people to be found to explain things to, you may end up in an endless loop, having a bot tell you the same annoying irrelevant thing over and over. How is that customer service?

In my own fantasy world, customer service is custom service—service that targets my needs and my problems, suggests good solutions, and assists me in implementing them, all via a human who can take into consideration the tangible and intangible contexts relevant to a remediation that fits my needs.

Sounds dreamy, doesn't it?

Such solutions actually exist in the real world. We have a store in Northampton that sells bras. When you go there, there are expert humans who work with you to find and fit bras that work for your body. I understand that this is an experience that half of the people around the world (and many of my readers) will never have, but trust me, it's wonderful. The experts at the store help triage a myriad of choices of bras: different colors, cup sizes, wired and unwired styles, sexy and utilitarian looks, different fabrics, and other options. They listen carefully as you tell them about the kinds of clothes you wear, your lifestyle, and your preferences. In the end, you can get something that really works for you and enjoy the experience of getting there. That experience is way different than choosing something on Amazon and hoping it will fit.

The bra store is all about custom service rather than homogenized, automated customer service. Custom service works there because the bra store is not a mega-company with millions of customers and we're probably a ways away anyway from automated systems that will find and fit your bra.

My experience with automated customer service is that it homogenizes individuals and the problems they are likely to have and separates them from the people who can actually help. When my wallet was stolen in New York City, FedEx was supposed to ship me a replacement card. There was no way to get a human from FedEx on the phone to deal with their temporary lapse of memory about where my apartment was. As far as I could tell, there were no humans to contact anywhere in the system except to go to an actual brick-and-mortar store, which was closed for a holiday weekend.

Look, I understand that it's harder and more expensive to do custom service at scale—when you have millions of customers—and that there's no cost-effective way to hire enough people to give everyone a personalized experience. But there's something else going on, something we should not lose in cyberspace: a real sense of respect for your customer and an understanding that without satisfied customers, businesses can cease to exist.

You can do custom service at scale with millions of customers. Some companies do this well and they invariably earn my loyalty. When I have a question about my Fidelity account that I can't answer by going to the website or chatting with the chatbot, I can call Fidelity. Within a relatively brief amount of time, the bot who answers the phone verifies my voice and connects me with a human who can track down the answer to my question. Something similar happens when I call USAA insurance. In 2025, USAA was in the Fortune 100 and Fidelity was the third largest mutual fund company in the United States.

Apple also has a phone number that connects me with real humans who can help solve problems. In addition, they have brick-and-mortar stores with a "Genius Bar" staffed by actual humans who apparently are not allowed to make me feel stupid. (OK, most of these folks are not geniuses but they understand Apple products and systems way better than I do.) You probably have your own list of companies that make it easy for you to deal with them and easy for you to solve problems. My guess is that all or most of them have knowledgeable humans available early and often.

Digital technologies should help us do things better, not worse. Algorithmic efficiency is not a substitute for human empathy and judgment. Can't we create a cyberspace that has both?
</rant> (end of rant)

The most egregious customer "service" systems are only the tip of the iceberg in a massive workforce shift that increasingly involves both humans and "intelligent" machines. Today's work at scale will need to integrate AI-driven work and human work responsibly to get the best of both. This brings us to the challenges and opportunities of the third increasingly prevalent, and profound, trend.

INCREASING "SMARTS": WILL AI TAKE MY JOB?

If you ask an AI researcher what the biggest challenge is for AI, they might give you a highly technical answer or perhaps posit about whether artificial general intelligence (the ability to think and learn like a human) or superintelligence (greater cognitive performance than humans) is possible. For the rest of us, "Will AI take my job?" is likely to be at or near the top of the list. According to the World Economic Forum in 2025, 22 percent of work tasks are performed mainly by technology, 47 percent are performed by humans alone, and 30 percent are performed by a combination of both. These are changing rapidly: "By 2030, employers expect these proportions to be nearly evenly split."[7]

Will AI take your job? Will it take mine? It depends. The answer is complicated and dependent on how we resolve critical challenges central to the way we work: When should technology and automation be used to enhance or replace human work (or humans themselves)? How do we balance the use of the tools at our disposal with our desire for human control? How do we empower society through the use of AI? Although these seem like new challenges, they are really the latest instances of questions we have grappled with before when powerful new technologies have come on the scene. Consider the Industrial Revolution.

The Industrial Revolution was a turning point in human history. Between the late eighteenth and early nineteenth centuries, innovations in iron production, chemical manufacturing, steam power, and other technologies reshaped the global economy and ultimately transformed society and the world. At their core, these new technologies expanded our *capacity for physical labor*, spurring tremendous innovation and enabling greater precision and scale. Industry after industry transitioned from hand production to machine production. People moved from rural areas to cities whose factories made new goods and services possible, and where jobs were plentiful and urbanization fed a growing middle class.

The Industrial Revolution transformed global economies, trade, commerce, agriculture, transportation, and lifestyles. New technologies created new jobs, replaced other jobs, and changed the character of many. The automation of physical labor shifted the nature of work and the workforce—we are still experiencing those societal impacts today. New modernizations also introduced many risks. In England, the majority of workers in factories springing up in new cities were young children, many of whom worked long hours in unsafe conditions. Many of these factories released large volumes of smoke and industrial chemical discharge, polluting the air. During his travels in 1835, Alexis de Tocqueville wrote of the pollution over Manchester, England, center of the cotton industry in the early nineteenth century: "A thick black smoke covers the city. The sun appears like a disc without any rays. In this semi-daylight, 300,000 people work ceaselessly."[8]

As risks to people and air quality mounted, public outcry helped spur the public sector to respond, and new laws and regulations were developed and passed. In the 1830s and 1840s, England enacted some of the first general laws restricting child labor, and by the 1860s, early pollution control laws—particularly targeting industrial air emissions—began to appear. These laws provided needed protections to reduce some of the most egregious risks brought on by the new technologies.

Fast forward to today. In our own time, advances in AI provide a means to expand society's *capacity for intellectual labor*. Like the opportunities to automate physical labor during the Industrial Revolution, today's opportunities for automating intellectual labor are beginning to transform work, the workforce, economics, trade, education, commerce, and many other areas. They are creating, changing, and replacing jobs. AI is driving new businesses and services and will necessitate new laws and governance. Like our predecessors during the Industrial Revolution, it will be incumbent on us to determine how to best use the automation of intellectual labor appropriately and in a way that leaves humans protected and in control.

Before we examine the potential of AI for intellectual work (yours, mine, and others) and in today's workforce, it's useful to note that while AI is an enormously powerful tool, *it's not the solution for everything*. As I write this, we are in a moment when the AI "hype curve" is stratospheric—seemingly everyone wants to jump on the AI bandwagon. AI is becoming the secret sauce in everything from credit scoring to compliance monitoring, toothbrushes to thermostats, whether it's actually adding value or not. (And for many products, conventional data analysis and automation techniques are now being reframed as "AI" to capitalize on this hype.)

This is precisely the moment when a public interest mindset can help us sort through the hype that promotes AI to human or superhuman levels. It can shape the way we think about AI and how we incorporate it into our work so that we leverage its best assets, use protections to minimize its most harmful risks, and maintain the human control necessary to get the best from powerful AI technologies.

So with a public interest tech mindset, we turn to the issue of whether AI will take your job. The answer will depend on what your job is, just what AI can do, and just what you can do.

Today, both the fastest-growing and fastest-declining job categories are driven by AI and digital technologies.[9] We need an ever-growing number of AI and data specialists and financial tech engineers, as well as software and application developers and autonomous and electric vehicle specialists. At the same time, we are losing clerks and cashiers, secretaries, bank tellers, and security guards, often to automation and surveillance technologies.[10] Despite an increasing focus on AI and data science in higher ed and an increasing number of students graduating with these skills, we don't have enough tech-trained workers. Many companies are focusing on tech learning and retraining on the job, an approach that works for some job categories but is too heavy a lift for others—sometimes exacerbating existing economic and demographic inequalities.

"Care jobs"—nursing professionals, social workers, counselors, and other caregivers—are also expected to see significant growth. So are higher education and secondary education teaching jobs.[11] Professional skills such as empathy and compassion, as well as a strong sense of ethics, are crucial for these jobs and cannot be satisfactorily replaced by AI. To succeed in all of these jobs in today's world, the training, retraining, and education of workers will also need to incorporate the responsible use and management of tech. This training must start early. In higher education? Definitely. In K–12 (lower grades)? Also yes! Gen Z and their younger sisters and brothers will grow up in a world in which AI is pervasive and makes many decisions that impact them. They will need to know how to responsibly use and manage these technologies.

As we consider how to incorporate AI within our work responsibly, it's worthwhile to pause and consider what AI does well and what it does not do well. What does the state-of-the-art in AI look like in the mid-2020s?

By 2026, AI has proven to be enormously helpful in triaging massive amounts of information and indicating what might be important, always with the caveat that high-risk AI recommendations and predictions must be vetted by a human in the loop. Lawyers, radiologists, and astronomers all use AI to sort through and analyze information from big instruments and massive libraries: Lawyers use AI to find cases establishing precedence and relevance to their own. Radiologists use AI to compare X-rays to massive image databases to identify tumors. Astronomers use AI to identify potential exoplanets (planets orbiting stars outside our solar system) by searching for anomalies in vast sky surveys. With the right prompts and human vetting of responses, the use of AI can be enormously labor-saving, helping to find needles in data haystacks.

AI can also help humans consider more options. Architects and civil engineers use AI to suggest new design concepts that combine aesthetics with structural integrity, efficiency, and sustainability. AI becomes a partner by generating new concepts for cutting-edge structures that align with project goals, location, environment, material properties, and other relevant information.[12] The AI-generated designs help human professionals think outside of the box and are a useful part of the process. New design concepts must be vetted by humans, making AI a partner that helps expand the horizons of human creativity and invention.

AI is also useful at subbing for humans when accuracy is *not* critical and in situations that are *not* high risk. Trying to find the next thing to watch on Netflix? The website Statista keeps track of the countries with the most titles available worldwide.[13] (In February 2025, it was Iceland with over 9,700 titles available.) Having a recommender system that sometimes (but not always) gets it right is helpful and low risk. Want to play bridge but don't have a fourth? Try playing with your threesome and AI. AI is widely used for data entry, coding, graphic design, and text editing. In many cases if the AI is wrong and the risks are minimal, the results are suboptimal, and everyone lives to see another day. However, higher-risk tasks *cannot* be left solely to AI.

Why not? At present, and likely for the future, there are things that AI *cannot* do. AI does not think, and experts disagree about whether AI will ever have the capability to reason as human beings do. Even if artificial general intelligence is possible, most agree that it is a ways off. Neither can AI feel. Stuart Russell, major AI researcher and coauthor of the authoritative text *Artificial Intelligence: A Modern Approach*, believes that our humanity is a critical differentiator in a workforce where decision-making is shared with AI.[14] Russell posits:

> When AI can do pretty much all the jobs that humans can do . . . the value that we can provide will be much more of an interpersonal nature . . . it's not going to be the value that a factory worker can provide because as we know, those types of jobs are already being automated out of existence. It's not going to be in routine clerical work . . . if you can no longer sell physical labor and you can no longer sell mental labor, it's not clear that there's another thing that the human race can fall back on except . . . interpersonal or emotional or empathic capabilities where we have this sort of intrinsic comparative advantage over machines because we know what it's like.

Russell continues,

> There's nothing a machine can do to find out what it's like. They can at best make empirical correlations . . . but they don't know what it's like. They don't know what it's like to be left by your lover. They don't know what it's like to lose a parent, or to lose a child, or to lose a job, or to be promoted, or any of the feelings of what it's like to be human. And so there we have this competitive advantage and there are also things we don't want to be done by machine.[15]

As Russell points out, today's AI is fundamentally limited. AI systems respond based on their training and learning; they can miss the subtle impacts of context and intangible nuances we all use to guide our impressions and thinking. They cannot consider things holistically and comprehensively as we do—factoring in personal experience, "reading the room," trusting our "spider sense." In helping my students use AI to aid (not do) their writing, I tell them to think of their AI chatbot as a flaky research assistant—one who might suggest ideas or ways of organizing material you hadn't thought of but who also is often inaccurate, tends to generalize, and doesn't have a particularly interesting writer's voice. (More on this in the appendix.) Everything an AI suggests should be checked (*don't trust* and do verify) as well as carefully scrutinized for relevance to the context and task at hand.

AI can be a great partner but is a poor leader. Because AI doesn't really think as a human does and doesn't feel, it doesn't really "know" what is ethical by *any*body's personal version of ethics. It can model empathy, but it can't be empathetic. It is not trustworthy—neither always accurate nor able to understand and make decisions about the real world based on all the information that humans use. So, will AI take your job? It depends on how your most human characteristics contribute to your work and, increasingly, on your ability to learn to use AI effectively as a tool. A public interest tech mindset indicates that in anything that involves real risk, humans must take the lead.

My own job as an educator, considered to be a "fast-growing" job by the World Economic Forum, is already forever changed by the power and peril of AI. My university colleagues and I will continually need to learn how to use AI effectively as a partner while helping students grow intellectually, emotionally, and as citizens and professionals who can succeed in the tech-driven world in which they (and we) live and will work. Educators are unlikely to lose their jobs, but our jobs are becoming different in fundamental ways, ways that make a responsible tech mindset even more important to our success.

In a real sense, to best manage the integration of AI into the workforce, we need to lean in to our humanity. Our sense of responsibility, empathy, ethics, and compassion are more important than ever. We should be working to ensure that our efforts in a complex world retain their richness and nuance and that our workforce retains its creativity and humanity, all parts of the fertile soil that has produced great artists, innovators, leaders, and thinkers throughout history. Today's world needs more of them, not fewer.

TOWARD A HYBRID SOCIETY

These three trends—surveillance, scale, and smarts—have begun to change the work we do and will have fundamental impacts on the future of both work and society. Not surprisingly, AI is beginning to drive all of them. As AI performs more of our work and has more decision-making autonomy, we increasingly live in a hybrid society—one in which both humans and machines share the responsibility for human well-being, planetary sustainability, and societal advancement, but where only humans are accountable.

This hybrid society has been the stuff of science fiction for many decades. It is now becoming our reality. How do we create a governance structure for this environment that maximizes the benefits of technology without creating "robot overlords" and increasing human risk? How do we ensure that humanity thrives? What are the lessons of history and governance that can help us deal with the challenges and opportunities of an AI-powered future? We turn to that next.

8

PUBLIC INTEREST IN A HYBRID WORLD: GOVERNANCE OF AI

When HAL 9000, the intelligent computer in the 1968 movie *2001: A Space Odyssey*, begins to malfunction in space and puts the mission at risk, astronauts Frank Poole and Dave Bowman discuss shutting down the computer. To prevent disconnection and continue the mission, HAL kills Poole by manipulating his air hose during a routine space maintenance task. HAL also disconnects the life support systems of hibernating crew members. Bowman, locked out of the ship by the computer, decides to disconnect HAL:

(Astronaut) Dave Bowman: Open the pod bay doors, HAL.
(Computer) HAL: I'm sorry, Dave. I'm afraid I can't do that.
HAL: . . . The mission is too important for me to allow you to jeopardize it. . . . I know that you and Frank were planning to disconnect me. And I'm afraid that's something I cannot allow to happen.
Dave: . . . All right HAL, I'll go through the emergency airlock.
HAL: . . . Just what do you think you're doing, Dave? Dave, I really think I'm entitled to an answer to that question. I know everything hasn't been quite right with me, but I assure you now, very confidently, that it's going to be alright again. I feel much better now. I really do.

Look, Dave. I can see you're really upset about this. . . . I know I've made some very poor decisions recently, but I can give you my complete assurance that my work will be back to normal. I've still got the greatest enthusiasm and confidence in the mission. And I want to help you. Dave. Stop. Stop, will you. Stop, Dave. Will you stop, Dave? Stop, Dave.

[Dave begins to shut down HAL]

HAL: I'm afraid. I'm afraid, Dave. Dave, my mind is going. I can feel it. I can feel it. My mind is going. There is no question about it. I can feel it. I can feel it. I can feel it. I'm afraid. I'm afraid.[1]

Today's sophistication about the promise and perils of AI (the movie was released well before many of today's AI luminaries were born) makes this conversation seem hopelessly naive. Yet now, in the mid-2020s, the specter of AI that is more capable than humans and that can (and will, according to some) prevail over human agency and decision-making haunts discussions about our ability to partner and coexist with intelligent technologies.

Today AI can read X-ray images, create art, beat human experts at Go, and pass an MBA exam at Wharton. It is an essential part of cyberspace, and AI-driven technologies are increasingly dominating the landscape. A senior exec at a Fortune 100 company told me that their CEO said that everyone needed a minor in AI. The top three corporate priorities? AI, AI, AI.

We increasingly live in a hybrid society guided by both humans and "smart," autonomous systems. And AI will be doing even more decision-making as technology and society evolve. Our legal and governance frameworks will need to adapt to this new hybrid world. We will need to make sure that humans can thrive.

How we think about AI and the relationship between humans and AI is fundamental to how we will govern in a hybrid world. How should decisions that affect humans be made and by whom? When should AI take the lead? Who is responsible (and accountable) when things go wrong? What happens when AI gets "better" and "smarter"?

In this chapter, we frame the challenges of creating governance for a hybrid world and human well-being. We start with a brief look at what artificial intelligence actually is. We then consider the relationship between AI and humans—a conceptualization that provides the foundation for any governance framework for a hybrid world. Key questions need to be explored about AI capability, liability, and power over humans. What are our options as we begin to conceptualize governance systems for a hybrid world? Let's explore.

WHAT IS AI?

Artificial intelligence is concerned with understanding and building intelligent machines that can compute how to act effectively and safely in a wide variety of situations. Intelligence itself (human and otherwise) is often defined imprecisely, but in general, subfields of AI seek to model human forms of intelligence including reasoning, planning, learning, and perception, as well as the ability to represent knowledge, process language, and act intelligently and autonomously.

In a classic text of the field, Stuart Russell and Peter Norvig observe that most subfields of AI occupy the spaces between *thinking* and *acting* and between *human* and *rational* behavior.[2] (For the mathematically minded, this defines a 2 × 2 grid.) This gives us four general categories (quadrants) with which to think about AI:

- *AI that acts like a human.* One metric of success for an AI that can act like a human is the *Turing test* (also known as the *imitation game*), where a machine passes the test if a human interrogator asking it questions cannot tell whether the answers came from a human or an AI. Subfields focusing on the capabilities useful for passing the Turing test include natural language processing, automated reasoning, knowledge representation, machine learning, computer vision, and robotics.

- *AI that thinks like a human.* Creating systems that think like a human requires that we understand the way humans think and can simulate this on a computer. Cognitive science explores how one might develop precise and testable theories of the human mind.

- *AI that thinks rationally.* Rational thought encompasses logical, evidence-based approaches to solving problems. Conventional logical systems assume certain knowledge about the world, but AI must solve problems in uncertain real-world situations. To manage uncertainty, researchers often incorporate probability into laws of thought models to simulate rigorous reasoning with imprecise or uncertain information.
- *AI that acts rationally.* We expect rational agents to operate autonomously, perceive their environment, adapt to change, and maximize their performance. They ideally create intelligent outcomes that "do the right thing," as defined by an objective function provided to the agent. Russell and Norvig observe that this approach has been a primary focus of AI research for most of the field's history. All of the subfields needed to pass a Turing test are needed to develop agents that act rationally. (Readers interested in an accessible and nuanced view of AI by experts in the field should read chapter 1 of Russell and Norvig's excellent text, which includes content on AI's opportunities, challenges, and history.[3])

For efforts in all of these categories, researchers use approaches from a wide variety of relevant disciplines. AI derives its methods and perspectives from philosophy, math, economics, neuroscience, psychology, control theory, computer science and computer engineering, among others. Today's AI (for example, the development of large language models (LLMs) like ChatGPT in machine learning) can also be resource hogs: Machine learning models require high-performance computers, immense amounts of data, tremendous energy usage, and a great deal of investment to reach the surprising accuracy and scope we are beginning to see. The level of funding needed for AI advancement has not always been available. Since the 1950s when AI was launched as an academic discipline, the field has gone through multiple cycles of optimism and funding, including "AI winters"—periods of decreased interest, funding and investment—and the current "AI spring," characterized by both high expectations and investment.

Experts in the field refer to the AI we see today as *narrow AI*—AI that solves specific problems like text generation ("write operating instructions for my vacuum cleaner in the style of Taylor Swift") or image generation ("create an image of the Pope in a puffer coat"). In a 2023 study of 2,778 AI researchers publishing in top-tier AI venues, respondents predicted that there was a 10 percent probability of "unaided machines outperforming humans in every possible task" by 2027 and a 50 percent probability of achieving this milestone by 2047 "if science continues undisrupted."[4] This suggests that many researchers believe that AI could reach the point where it is comparable to general human intelligence (artificial general intelligence, or AGI). Some believe it can become even more capable than human intelligence (superintelligence). Will the ability to perform human tasks be enough? Will AI be able to simulate the experience and human judgment needed to assess context and perform tasks in ways that produce positive impact without introducing human harm? The jury is out.

But already AI is everywhere: Machine intelligence is embedded in your search engine, your insurance company, your social media, your government services, your law enforcement agency, and your financial system. The notes from your last doctor's visit on your patient portal may have been generated by AI. AI-driven facial recognition systems routinely provide evidence with which you can be charged for a crime. In many cases the outputs of AI are good enough, but in other cases they put humans at risk. Will "better" AI, AGI, or superintelligence improve things? It's too soon to tell. Whatever the future holds, the decreasing gap between AI outcomes and human behavior indicates that AI has the power to change the present and future trajectory of humanity.

How will we move toward our AI future? The list of areas where next steps are needed is long: As AI is increasingly embedded in autonomous systems—cars, robots, and other objects—the quality, cost, and capacity of sensors and actuators will need to improve to create the autonomous AI agents of tomorrow. Further research on the underlying mathematical models, engineering, and software systems for AI is required to solve many of the problems AI researchers find difficult today. Running AI models—cost-intensive and energy-intensive on today's computers—will drive next-generation hardware innovations. We will also need to rethink the economics of AI, and especially the intense resource requirements of today's LLMs. All of this means *it will take time* to reach the potential of AI (and possibly AGI) and its role in our future, whatever that is.

Most importantly, future innovations in AI will depend on the innovators. Russell and Norvig remind us that AI is human-made and "*we* [humans] design AI systems, so if they do end up 'taking control' . . . it would be the result of a [human] design failure," meaning that if we end up with evil AI overlords, it's our own fault.[5] Tech responsibility must drive the innovation, creation, use, and management of AI systems and services so that they ultimately empower humans to thrive.

Many others reinforce the fundamental role of humans in creating AI's trajectory. Political scientist Josh Simons points out that AI, like all technology, is inherently biased, political, and reflective of the attitudes, assumptions, cultural references, and values of humans: "Although they appear purely technical, choices involved in machine learning, by prioritizing the interests of some social groups over others and protecting some fundamental values while violating others, raise fundamental questions about the purpose of decision making. Machine learning is political."[6] Cathy O'Neil reinforces this point in her iconic book *Weapons of Math Destruction* which describes multiple examples of data-driven and AI decision-making that put humans at risk: "models, despite their reputation for impartiality, reflect goals and ideology . . . Our own values and desires influence our choices, from the data we choose to collect to the questions we ask. Models are opinions embedded in mathematics."[7]

The AI we create today and in the future has a direct impact on the way we experience our world. It may (or may not) be more competent or accurate than humans. It *will* incorporate human biases and worldviews without the comprehensiveness of human experience. The governance systems for this hybrid world will be based on working models of the capabilities of its decision-makers and about their relationships to one another. We have a good idea and much experience with the characteristics and capabilities of humans. We also need a reasonable model of what we can expect from AI.

WHAT IS AI CAPABLE OF?

- You're at Thanksgiving dinner talking to Cousin Damien about hiring for your business. Damien says you should favor applications from candidates who went to coed colleges over all-women's colleges, pointing out that your current successful employees went to coed institutions. Is this discriminatory?
- You're talking with your friend Brenda. You tell her you feel like a failure and your life has little meaning. Brenda tells you that she's really sorry to hear that and offers you tips on being more productive. Is this sensitive?
- Your dad is doing great, but sometimes he makes mistakes. He's driving through a residential area with a stop sign at an intersection. The stop sign has some graffiti. Your dad mistakes the stop sign for a speed limit and barrels through the intersection at 45 miles per hour. Should you trust his driving?

Many of us would dismiss Damien's response as biased, Brenda's response as callous, and your dad's driving as foolhardy (and maybe a sign that it's time to take away the keys). All of these are clearly situations where humans need to act rationally, ethically, empathetically, and legally. Yet these are all situations that can or have happened when AI (not Damien, Brenda, or your dad) was in charge.

As "smart" as AI becomes, will it ever have real capacity to think, feel, and act morally and compassionately? Will it ever reliably and adequately simulate the behavior of a smart, empathetic, ethical human? Today's AI can simulate some human behaviors but it does not have full human capabilities. It can pattern-match but it cannot think. It can simulate emotion but it cannot feel. It can advocate but it does not believe. It has no moral compass. Any ethical behavior is simulated based on what AI has inferred from what it has been fed.

Experts disagree about what AI will ultimately be capable of. In 2021, AI researchers Timnit Gebru, Emily Bender, and colleagues described LLMs like ChatGPT as "stochastic parrots"—sophisticated statistical models that can make surprisingly accurate complex predictions in response to prompts (inputs) but which don't really "know" what they are doing: "Contrary to how it may seem when we observe its output, an LM [language model] is a system for haphazardly stitching together sequences of linguistic forms it has observed in its vast training data, according to probabilistic information about how they combine, but without any reference to meaning: a stochastic parrot."[8]

Less than five years later (an entirely new generation in AI's rapid technological development), many experts still concur. "The technology we're building today is not sufficient to get [to AGI]. . . . What we are building now are things that take in words and predict the next most likely word, or they take in pixels and predict the next most likely pixel. That's very different from what you and I do," comments Nick Frosst, formerly of Google and founder of the AI startup Cohere.[9]

Bill Peebles, a lead researcher for video generation at OpenAI (the company that created ChatGPT) sees AI as having more capability: "As long as you can get every piece of information in the world into these models, that should be sufficient for them to build models of physics, for them to learn how to reason like humans."[10] Peebles believes that the current trajectory of machine learning could give rise to intelligence; in the end what matters is the pursuit of data.

Stuart Russell, coauthor of a classic text on AI, sees AI as limited in that it will never have the capacity to "know what it's like." As we saw in chapter 7, Russell believes that the capacity for emotion and empathy is an intrinsic comparative advantage for humans, that "there's nothing a machine can do to find out what it's like. They can at best make empirical correlations."[11]

However the controversies play out around whether AI will ultimately be capable of thinking like humans, feeling, or determining right from wrong, experts agree that we are a long way from a comprehensive understanding of the potential capabilities of AI. The way we govern AI should focus on what AI is rather than what we hope or imagine it will be. *A good governance framework should protect humans against the AI risks we experience in the present and be able to evolve dynamically to protect us from the potential risks of AI in the future.*

The lowest-risk option for any governance system is to leave high-risk decisions and risk-prone analyses to humans and to find a way for there to be a human in the loop to supervise, partner, check, and vet the outcomes of AI on which humans depend. How should a hybrid governance model conceptualize and encode the desired relationship between humans and AI? Should we be equals—in status, responsibility, liability, priority, something else?

WHAT SHOULD THE RELATIONSHIP BETWEEN HUMANS AND AI BE?

How we conceptualize humans' relationship with AI—with respect to law, regulation, practice, and enforcement—lays the foundation for hybrid governance systems and their ultimate impact on humans. What are viable models for this relationship? Science fiction provides us with a lot of examples of possible relationships between humans and AI. From Mary Shelley's *Frankenstein* to the evil robots in *Terminator* to the romantic relationship between a lonely human and a smart operating system in *Her*, we have a host of scenarios for how relationships between humans and AI could play out.

One of the most famous descriptions of how we might conceptualize the relationship between humans and robots/AI comes from Isaac Asimov, writer and biochemist. In his iconic 1942 short story "Runaround," Asimov created three "laws" for engagement between humans and robots (and subsequently added a "zeroth" law preceding the others in importance):[12]

0. "A robot may not injure humanity, or through inaction, allow humanity to come to harm."
1. "A robot may not injure a human being or, through inaction, allow a human being to come to harm."
2. "A robot must obey the orders given it by human beings except where such orders would conflict with the First Law."
3. "A robot must protect its own existence as long as such protection does not conflict with the First or Second Law."

Clearly, the computer HAL 9000 in *2001: A Space Odyssey* violated all four of Asimov's laws. (Under a useful governance framework for a hybrid world, we would have some idea of what legally should happen next and who would be accountable.)

Roboticists don't take Asimov's laws very seriously—they are too abstract and high-level. They are also insufficient for addressing today's challenges with AI. Like laws that promote high-level aspirations such as world peace or equality for all, the devil is always in the details. And there are a LOT of details needed to make Asimov's laws actionable.

In the interest of specificity and creating robot governance more targeted to today's AI challenges, law professor Frank Pasquale adds four extra laws to Asimov's list:[13]

4. "Robotic systems and AI should complement [human] professionals, not replace them."
5. "Robotic systems and AI should not counterfeit [recreate] humanity."
6. "Robotic systems and AI should not intensify zero-sum arms races."
7. "Robotic systems and AI must always indicate the identity of their creator(s), controller(s), and owner(s)."

Pasquale's additional laws provide guidance for a hybrid world in which robots/AI *complement* rather than replace humans. We saw this prevalent theme when we considered AI "smarts" in the workforce in chapter 7. Most importantly, a focus on AI as an enhancement for human efforts argues for a governance system that is *human-dominant* and incorporates a human in the loop when AI is making high-risk decisions. But specifically, how do we do that? What would a human-dominant body of law, policy, and practice look like? One that recognizes the benefits of AIs as autonomous agents and their capacity to enhance human efforts while providing needed protections against undue risk.

Roboticist Kate Darling suggests that we get closer to an effective and actionable model for the human/AI relationship if we consider the relationship between humans and *animals*.[14] Hear her out: We have thousands of years of rich and nuanced experience engaging with animals—as pets and companions, as workers (horses, oxen, sled dogs), as weapons (attack animals), as food, domesticated and in the wild. We assume that animals have autonomy, can reason, and can collaborate with humans. We don't assume that animals have a moral compass. We don't assume that they think the same way we do. We coexist with them and have well-defined practices and rules for a broad spectrum of situations in which we engage. Rules that cover behavior, liability, and even rights.

With respect to governance, sometimes we think of animals as property (e.g., lab rats) and often we think of them anthropomorphically (we love our pets and believe they love us in return). We have a human-dominant set of laws that frame how we engage with animals but also give animals some "rights" (think laws against animal cruelty or protection of endangered species). The variety of human/animal governance frameworks around the world also reflects differences in how communities, cultures, and religions view animals, and may differ with respect to animal treatment, protection, and desired role within society and the biodiversity ecosystem.

All of this is complex, nuanced, and covers a broad spectrum of scenarios. But ultimately the way we conceptualize relationships between animals and humans in governance allows us to manage liability, responsibility, and risk in a way that recognizes the autonomy and independent (and different) intelligence of animals while still promoting human well-being. Darling suggests we leverage this rich body of law and experience in defining our relationship with robots—a "new breed"—and creating governance structures in which AI and humans can coexist. In her book on the subject, she provides a wealth of examples that help guide us in thinking about governance of this new breed that, like animals, is sometimes treated as property, sometimes as people, and sometimes as something in between.[15]

We are most familiar with treating AI systems the same way we treat other technology—as property—with responsibility for harm often falling under the general category of product liability where the company or some representative(s) of it is held accountable. In general, this works when a product is badly designed or faulty in some way and the company is responsible. But AI can make *independent* decisions that harm people without "understanding" their impact. Who should be responsible when a self-driving car misclassifies a pedestrian and hits them? Who should be responsible when the AI system that causes harm is open-source and there are many contributors? Conventional product liability law doesn't cover these situations very well.

Using human/animal relationships as a model for human/AI relationships expands our options. Responsibility for AI harm may be hard to assess but *accountability can only be in the human realm.* Darling notes that "the animal analogy helps us start from the baseline that robots don't make decisions like humans and that the responsibility for the harm needs to be assigned to people."[16] Remediations for negative outcomes must address the harm but do not always need to specify the blame.

Human/animal relationships also give us multiple ways to deal with accountability. In the animal world, pets may be regulated through their owners, who can be held accountable for their behavior. In some situations, owners may need to license their dogs, leash and muzzle them if they are considered dangerous, or go through mandatory training. An AI version of this might involve requirements for companies to mitigate the greatest potential harms from AI by restricting use of these systems for high-risk human or ethical decisions. This would put the onus on providers to employ responsible design, use, and management controls. The animal world sometimes imposes insurance requirements on owners of dangerous animals—funds that help compensate for harm when it does occur without specifying who or what is to blame.[17] This provides a no-fault mechanism for dealing with compensation. (We already do this with cars in some areas and are likely to see more of this with AI-driven cars.)

Treating AI as "people" with rights and responsibilities under the law is a nonstarter. But in the animal world, it has been done (rarely). Darling cites the 1457 murder of a five-year-old boy in Savigny-sur-Étang and its trial that ended in the hanging of the murderer—a pig.[18] This seems far-fetched from a twenty-first century perspective yet the specter of harm done by AI to humans is very real.

Darling's human/animal analogy suggests that we shouldn't conceptualize all our interactions with AI with a broad brush and that AI should not be allowed to dominate in scenarios that provide real risks to humans. Effective human-dominated governance systems can support this nuance and contextualization of our relationship with AI. This gives us a way to create a new conceptualization of human engagement with AI, one that leverages AI's value to us as a tool, device, partner, collaborator, and sometimes companion without expecting it to think as we do, understand, or know right from wrong.

GOVERNANCE OF AI-DRIVEN CRITICAL INFRASTRUCTURE

Governance is particularly important for digital critical infrastructure where AI increasingly takes the lead. Political scientist Josh Simons observes that such technologies are a special kind of critical infrastructure, one with the capacity to *change* society, exemplified by Google Search and Facebook: "Facebook's and Google's machine learning systems have become part of the infrastructure of the digital public sphere, shaping how citizens engage with one another, access information, organize to drive change, and make collective decisions. Their unilateral control over these ranking systems involves a distinctive kind of infrastructural power. Unlike railroads or electricity cables, Facebook's newsfeed and Google's search results not only enable people to do what they want to do but shape what people want to do."[19]

Because of AI's capacity to shape our experience in the public sphere, Simons proposes public governance of essential AI technologies. In particular, he proposes that companies overseeing these technologies be considered as *democratic utilities*, a new category of corporation "whose infrastructural power shapes the possibilities of collective self-government itself."[20]

Democratic utilities may have been born in the private sector but must be managed in the public sector. Simons cites Facebook and Google as examples of the type of company that should be governed as democratic utilities: "[Google and Facebook's] unilateral control over [their] ranking systems concentrates social and political as well as economic power, shaping how we understand and interpret the world around us, discuss fundamental matters of common concern, organize social and political groups, and make collective choices."[21] Simons reinforces a key message of this book: The public sector must oversee essential information technologies as if they were critical infrastructure because, in fact, they are. AI-driven essential technologies need to be *managed to promote the public interest*, and *companies/people need to be held accountable* when public risks and harms are unacceptable.

As democratic utilities, AI-driven essential technologies could be regulated by a public agency—Simons calls it an "AI Platforms Agency (APA)." The purpose of the APA would be to develop and deploy mechanisms of participatory governance—involving civil society groups, public bodies, and citizens—to evaluate how the design and system metrics of essential technology platforms support the public sphere.

Democratic utilities should be accountable for supporting the social and informational conditions for healthy collective self-government. Simons suggests that the APA could assess democratic utilities based on how they prioritize the public interest over advertising revenues (which he labels *anticorruption*), promote a multiplicity of distinct voices and values in the public sphere (labeled *diversity*), and support a common *shared experience* via a civic information architecture that supports broad dissemination of public information.[22] His approach suggests a new way to fashion a regulatory landscape for digital critical infrastructure, one more public interest–focused than the market-influenced approach to governance used today.

Another way to govern AI, suggested by legal scholar Tim Wu, is to structure essential technologies using a "separations principle" for corporate information monopolies that traffic in forms of individual expression.[23] Both Simons and Wu suggest that essential technology companies (democratic utilities) can create distance or firewalls between major system functions on their platforms, for example structurally separating Google Search from Google Ads. This could allow Google and other companies to harmonize business goals and public interest goals more effectively.[24]

Ultimately, the point is that humans created these systems, complicated as they are, and have guided the most prevalent of them to become enormously profitable and successful in the marketplace. Companies (and their humans) have the ability to reframe these systems to become more public interest–focused, especially when required to do so by a public sector agency with enforcement powers. All of this must happen in the context of a governance system driven by the public sector; it will not be done voluntarily by the private sector.

EMERGING AI GOVERNANCE TRENDS

So what should a governance framework for a hybrid world look like? We are too early in the game to have mature examples, but early experiences with developing principles, legislation, and interpretation of existing laws, regulations, and policies point to three important trends, described below.

Trend 1: Leverage Existing Law, Regulation, and Policy

As disruptive as today's applications of AI are, it is important to remember that they do not exist in a legal vacuum. All over the world, there is a patchwork of laws, regulations, policies, and industry practices that apply to technology's design, development, deployment, use, and management.

Before the accusation of AI "plagiarism" in the training of ChatGPT (as per the *New York Times* lawsuit against OpenAI and Microsoft), there was copyright law. Before the American Civil Liberties Union's privacy intrusion suit against Clearview AI (a facial recognition company), there was biometric privacy legislation in the state of Illinois.[25] We are now undergoing a worldwide experiment to interpret and apply existing legal structures to AI. We are not starting from scratch.

As we saw in chapter 5, laws and regulations are usually only as efficient as their enforcement mechanisms. Enforcement of the state of Vermont's Data Broker Law was challenging because of small fines and a focus on registering rather than neutralizing the risks of data brokers operating in Vermont. Enforcement of the European Union's General Data Protection Regulation was more effective, thanks in part to adequate administrative infrastructure and high fines for noncompliant companies, reaching into the millions and billions of euros.

Today, we are seeking to leverage all kinds of existing legal controls, including copyright, antimonopoly, privacy, security, product liability and other legislation and regulations to guide and manage AI toward the public good. This approach treats AI as property and is challenged when AI makes independent "decisions," but a growing cache of AI-relevant legal interpretations and case law is providing part of the foundation for a new era of AI governance.

Trend 2: New Outcomes-Oriented AI Legislation and Policy

Existing law with new interpretations is a good start but won't be enough; it is necessary but not sufficient. Nations and jurisdictions all over the world are preparing themselves to support a public-focused (and human-dominant) legal structure for AI through the development of new principles, policies, and nascent legislation. These new laws, regulations, policies, and practices will need to be created in a dynamically changing landscape where advances in AI innovation, public opinion about AI, and the dramatic uptake of AI within the private sector are all evolving rapidly.

Because of the dynamically changing landscape in which hybrid governance systems will be developed and evolved, there are more discussions, principles, guidelines, and studies than there are laws and regulations at present. Jurisdictions all over the world are studying how AI should fit within existing national systems of commerce, resource management, public protection, law, and policy and what new legislation and policy is needed. Arguably most advanced in terms of protections is the EU's AI Act (chapter 4), whose rules go into effect in 2026 and 2027. It is also a model of why successful legislation should not be product-specific. Legislation that focuses on ChatGPT specifically, for example, may not apply to AI systems Claude, CoPilot, Gemini, DeepSeek, and other commercial alternatives. The genius approach of the AI Act's coupling of risk with public scrutiny helps avoid this issue.

Political regime change can also make a difference with respect to which new legislation is created and how existing legislation is enforced. In chapter 4, we described President Biden's 2023 executive order "Safe, Secure, and Trustworthy Development and Use of Artificial Intelligence" and its revocation and replacement in 2025 with President Trump's executive order "Removing Barriers to American Leadership in Artificial Intelligence."[26] As a consequence, the focus of AI policy and law in the United States has shifted from public protections to AI primacy. Which public protections and primacy strategies will be promoted by the US federal government as part of the next generation of law and policy is still to be determined.

In dealing with the disruptive effects of AI and developing new legislation, it is important that humans do not see themselves as victims. In legal scholar Nathalie Smuha's excellent book, *The Cambridge Handbook of the Law, Ethics and Policy of Artificial Intelligence*, a compendium of thoughtful essays on the state-of-the-art of AI governance circa 2025, she reinforces a theme we have advocated for throughout this book: "The idea that AI overcomes us, and that all we can do as a society is run behind the facts and try to come up with some imperfect legal rules to control it, is both erroneous and dangerous, as it unwittingly deprives us from our individual, collective and societal self-determination vis-a-vis technology, which remains a human made and human caused object."[27]

A well-worn theme in science fiction that has *not* gained real traction in governance discussions is whether robots should have "rights" that potentially conflict with human rights. A focus on human rights over potential rights for robots is reinforced by Darling's approach of using human/animal relationships as a model for human/robot relationships. Is there any scenario in which robots might be afforded "rights" over humans in a hybrid governance structure? Many legal scholars say no. We expect entities with rights and privileges to be autonomous, independent, and accountable. AI systems have the first characteristic, but even (future) self-driving cars will not be independent of the internet and infrastructure that must feed them data for operation. AI systems for the foreseeable future have neither the "understanding" nor the capabilities needed to be accountable. Law professor Kiel Brennan-Marquez points out that it would be hard to attribute legal rights to an entity that "does not have a life of its own."[28]

This perspective is reinforced in an intriguing paper by Abeba Birhane, Jelle van Dijk, and Frank Pasquale called "Debunking Robot Rights Metaphysically, Ethically, and Legally." The authors give multiple reasons why the notion of "rights" for AI/robots does not hold up under traditional and structural approaches:

> *Metaphysically* . . . machines are not the kinds of things that may be denied or granted rights. . . . *Ethically* . . . given machines' current and potential harms to the most marginalized in society, limits on (rather than rights for) machines should be at the centre of current AI ethics debate. From a *legal* perspective, the best analogy to robot rights is not human rights but corporate rights, a highly controversial concept whose most important effect has been the undermining of worker, consumer, and voter rights by advancing the power of capital to exercise outsized influence on politics and law.[29]

Until AI can master agency, autonomy, independence, accountability, and moral consideration, real rights seem off the table.

Trend 3: Law Is Not the Only Tool

Part II of this book demonstrates that there are many ways to responsibly influence tech, including AI, apart from legislation and regulation. Designing, developing, deploying, and using AI responsibly are all complementary to legal and regulatory structures focused on managing tech responsibly. We promote responsibility and apply a public interest tech mindset when we build (almost) self-driving cars that require a human to be in the loop to oversee vehicle operation and respond to high-risk situations, when we develop AI systems on training sets that better represent the people they impact, or when we limit the environments in which AI will be used to lower the risk of harm.

As Smuha points out, "we should not put all our hope on new or better legislation, or on legislation altogether. . . . No law is perfect, as it is always a compromise solution that must meet the political reality, and that has inherent tensions between the demands of generality and specificity, and between predictability and flexibility. Rather than losing faith in the power of the law altogether, this should instead incite us to also consider how law complements and interacts with solutions outside the legal realm."[30]

However we evolve current systems of governance to meet the challenges of a hybrid world, we will need to find pragmatic, effective ways to reap the many benefits of AI while limiting its serious risks. How close are we to comprehensive and effective governance systems? Really, pretty far. We need to have a better idea of what AI is capable of and what our relationship with it should be. We need to understand its risks and potential mitigations in multiple scenarios. In tomorrow's governance systems, as in today's, one size will not fit all and we are likely to interpret existing law and create AI-relevant law, regulation, and policy for different situations piecewise—industry by industry, common scenario by common scenario, constituency by constituency, nation by nation.

Still, we need to be thinking about all these things *now* to avoid a world where AI makes decisions largely unbridled by public management and governance structures that can mitigate potential harm. We need to be ready for peaceful and beneficial coexistence with AI—it is not too soon to start thinking about how we should accomplish this.

WHEN SHOULD AI TAKE THE LEAD? THE AI "SERENITY PRAYER"

(Hopefully) most of us agree that AI should not be the dominant decision-maker in situations that unduly put humans at risk. But we also shouldn't throw out the baby with the bathwater. AI currently takes the lead in a variety of areas that humans find both convenient and useful—areas that free us up to do things that better leverage our uniquely human capabilities. AI does handwriting recognition for the post office, traffic management in complex environments, cancer tumor triage in X-ray images, and a host of other things that, for the most part, we are happy to cede to "smart" machines.

More generally, AI is tremendously useful when its results are "good enough" and when humans are not put at serious risk when they are off the mark. In practice, "good enough" means safe enough, private enough, fair enough, and so on. In other words, "good enough" means that AI promotes public interests adequately enough to foster societal well-being and avoid too much harm. As AI gets "smarter" and better, when should we cede decision-making to AI? That's a question for both the present and the future.

We've already identified the path to a good answer—the human in the loop. Until we've thoroughly vetted new AI capabilities and tools, it's critical that humans make the ultimate decisions—that they serve as pilots to AI's copilot. Humans (also not infallible . . .) are the only ones who can truly be accountable for risk. That means that humans need to assess harm and be responsible for pressing the GO button for high-risk AI decisions. Once we have enough experience and understand the risks of AI tools and services thoroughly, we can determine whether they can operate on their own. Until then, it's humanity's responsibility to promote a positive outcome.

In the end, we will ultimately need to create our own AI serenity prayer: *Grant us the knowledge to assess the scenarios for which AI is good enough, the scenarios for which humans must make the call, and the wisdom to know the difference.**

How do we achieve this wisdom? It begins with good governance. But to truly thrive, we must reimagine our role in cyberspace. We explore that in the next and final chapter of this book.

* The "usual" Serenity Prayer is: God, grant me the serenity to accept the things I cannot change, the courage to change the things I can, and the wisdom to know the difference.

9

RETHINKING CYBERSPACE TO PUT PEOPLE FIRST

Cyberspace is a work in progress, created every day by you and me as we click, search, post, code, view, and otherwise engage with digital technologies. We continuously create cyberspace, and in some sense, cyberspace continuously creates us. My goal in this book is to help readers develop a public interest tech mindset and to provide pragmatic strategies we can use to nudge cyberspace closer to promoting the health and well-being of humans and the societies in which we live.

In part I, we considered the changing role of cyberspace in our lives and the importance and growing risks of its essential technologies, our digital critical infrastructure. Like the nondigital world, when infrastructure becomes essential, the public *must* become a prime priority in its creation, use, and management. Risk must be mitigated and reduced, infrastructure must serve everyone, society must benefit. In part II we focused on different strategies to reduce risk for digital critical infrastructure—for users, policymakers, developers, educators, communities, companies, and organizations. Reducing risk is contextual, depending on many specifics of the situation, people, and organizations involved. Successful strategies leverage these specifics to achieve positive benefit. A public interest technology mindset serves as an effective way to think about reducing risk, a way to identify and deploy strategies that promote better tech and tech that is better for us.

Sometimes it helps to consider things from a greater distance, where it's harder to see details but easier to get the big picture, to differentiate trends from noise, as the statisticians might say. What is the big picture of cyberspace—an ecosystem of interdependent technologies and people that has begun to fundamentally change the way we work and live? I end this book with two musings about the big picture of cyberspace, different ways of looking at cyberspace's "residents" and at its invasive technologies that may help us develop additional approaches to nudge cyberspace to be healthier and more empowering for us all.

The first musing focuses on the *ethnology* of cyberspace—how should we consider the "residents" of cyberspace—humans, users, and bots? What is the difference between them and how does it impact the way we engage with and manage the digital world? The second musing focuses on the *ecosystem* of cyberspace—the invasive potential of its essential technologies—and considers how we might approach risk reduction using an intriguing analogy from cancer research.

We start with the residents of cyberspace—the entities who shape our experience in the digital world.

MUSING 1: RETHINKING THE ETHNOLOGY OF CYBERSPACE

In the beginning, there were always the users. Ever since the invention of the internet and the creation of its products and services, the infinite variety and diversity of humans have been (and realistically must be) represented finitely—as *users*, circumscribable entities associated with a specific set of interests and descriptive characteristics. The organization Cambridge Analytica, famous for collecting and using data from Facebook users to potentially skew the 2016 US presidential election, used over 4,000 data points to identify each user.[1] That number is no doubt child's play compared to the data that Google, Amazon, TikTok, and other companies scrape, collect, share, and buy to represent each of their users, trade secrets all.

User representations are not just flat aggregates of characteristics. Sophisticated user profiles group human characteristics into aggregate representations that seek to model interests, belief systems, and behaviors. Cambridge Analytica used psychographic profiles based on demographics, data analytics, and behavioral characteristics—each individual was characterized in terms of their "openness," "conscientiousness," "extroversion," "agreeableness" (likelihood of putting other people first), and "neuroticism" (how much they worry), all derived from Facebook and other data. At a 2016 conference, Alexander Nix, then head of Cambridge Analytica, chillingly described how psychographic user profiles were paired with persuasive targeted messaging. To promote gun rights, for example, "conscientious, neurotic" individuals would receive messages about the importance of the right to bear arms under threat of a burglary; "closed, agreeable" individuals would receive messages that gun rights are patriotic and part of the family tradition between fathers and sons.[2] The goal of psychographic messaging during the election? Create many paths to the same destination—support for the candidate paying the bill.

Cambridge Analytica is only the tip of the iceberg. There is no doubt that user profiles are getting more sophisticated and better at modeling human behavior and our inclination to engage. Fed by increasing amounts of data, the gap between our user profiles and our human preferences and behavior is narrowing. For all practical purposes, does it matter anymore whether I am a human in cyberspace? Is my representation as a user good enough?

A lot of the time, *it is*. For the most part, companies don't really need to know me in all of my complexity to provide useful services. Amazon "knows" I am a kayaker (or at least markets to me that way). It doesn't know how I feel when I'm on the water nor does it know whether I can do an eskimo roll (a way to right a capsized kayak). It doesn't really have to. The shopping algorithms feed on my data and follow my lead when

I look at paddles or gear, keying off my own inclinations. Honestly, for kayaking, that's convenient and good enough.

But sometimes, my user representation is not good enough and I need my cyberspace services to deal with the complexity of me as a human—my inconsistencies, my sensitivities, my sense of ethics, my evolving experience with and understanding of the world. Automated customer service systems are often good at dealing with users and bad at dealing with humans—they focus on common user problems that can be summarized in frequently asked questions—common cases but not edge cases. They leave out a boatload of problems whose nuanced details live in the gap between users and humans.

Many of humans' biggest challenges in cyberspace—trustworthiness and authenticity, access to opportunity, fair and compassionate treatment—live in that gap. We are complex and multidimensional, but our technological user profiles are profoundly reductionist. They are essentially sophisticated cartoon representations with insufficient capability to include belief, desire, experience, the ability to learn and change our minds, and all the other parts of us that are hard to quantify and monitor. Targeting us as users without recognition of our humanity *asks us to adapt to cyberspace*—to fit into a limited representation of our interests and characteristics. Targeting us as humans would *force cyberspace systems to adapt to us* by providing options to better deal with context, and by recognizing our autonomy, ability to exert control, and infinite capacity for human emotion, experience, and unpredictability.

Narrowing the gap between humans and users is the ultimate brass ring for digital goods and services. Many in the private sector believe that if we could just capture *everything* that influences human behavior and decision-making, we might close that gap. That approach requires both an unimaginable amount of data and confidence that digital systems can somehow organize and simulate the fulsome complexity of human experience and behavior.

I believe that closing that gap is a fool's errand. Better to acknowledge that we will always be represented as users in cyberspace and that we must design and manage technology for more human control, autonomy, and respect, and provide space for our many exceptions. Better to recognize that human behavior is complex and stems from experience beyond what can be represented by our profiles.

The academic and commercial area of human–computer interaction (HCI) focuses on how to improve human engagement with computers, in effect exploring the gap between humans and users. The field is not new. Don Norman is a cognitive scientist who wrote *The Design of Everyday Things* (originally titled *The Psychology of Everyday Things* in 1988), which many consider to be a bible of HCI.[3] He joined Apple in the 1990s as "user experience architect," the first person to have "user experience" as part of their title.[4] In the 2000s, Norman began to use "people" instead of "user" as the target of design, commenting, "People are rich, complex beings. They use our devices with specific goals, motives, and agendas. Often they work with—or against—others. A label such as customer, consumer or user ignores this rich structure of abilities, motives, and social structures. Time to admit that we are people, that we design for people."[5]

The design and consulting firm IDEO and Stanford's d.school have done much to popularize human-centered design and design thinking as people-centered approaches to innovation and problem-solving. These approaches focus on empathy, ideation, prototyping, and collaboration. Like the people-centered methods described in part II of this book, human-centered design and design thinking focus on the gap between user and human. They are the design and development yin to responsible tech management's yang—the societal controls needed to ensure that the public interest is a priority in technology deployment and use. Both design and management—yin and yang—are needed to ensure that human experience is elevated in cyberspace.

Beyond humans and users, other "life forms"—bots—are increasingly taking up residence in cyberspace. Bots are software applications that automate tasks on the internet. They are software-sourced agents that can be designed to operate independently as users—sometimes for "good," and sometimes for "evil." Bots span the spectrum: Some bots crawl the web to scan and index web pages—behind-the-scenes work critical for search engines. Some bots scan prices to look for bargains on Amazon, eBay, or in the stock market. Some bots turn computers into "zombie machines" that can be coordinated as a group—a botnet—to automate the distribution of spam or participate in a denial-of-service attack that overloads a website's servers with traffic. Some high-profile bots are AI-driven software agents like ChatGPT, bots whose rapid rise is impacting both the ethnology of cyberspace and our use of planetary resources.

A cyberspace whose residents are users, humans, and bots is a complicated thing. One that must be managed proactively to put people first. It is not in humans' best interest for bots to "gang up" to skew the prioritization of information, who gets heard on social media, or which product is considered best. We now proactively use some approaches to ensure that a user is human-sourced rather than software-sourced: For example, CAPTCHA (Completely Automated Public Turing test to tell Computers and Humans Apart) is a step you often see between login and service provision that asks you to identify bridges, motorcycles, words, and other things—all to determine that you are a human-sourced user and not a bot.

The colonization of cyberspace by AI will increase the complexity of cyberspace ethnology. How will we deal with the interaction between humans, users, and bots? Current and emerging examples are context-dependent but provide some hints. In the transportation sector, we are moving toward an environment where self-driving cars will be prevalent. Engagement on our roads will move from human driver–to–human driver to AI-to-AI (bot-to-bot) interaction as platoons of self-driving cars communicate with each other and AI-powered traffic systems. What active role will humans play in this landscape? How should we expand safety strategies to minimize new sources of accidents (e.g., denial-of-service botnets)? As we plan transportation systems of the future, we must begin in the present to think through their potential for human harm and not just wait for problems to occur.

In social media, we are already dealing with bot-to-bot engagement where bots can dominate discussions, skew public opinion, and sometimes create mob behavior for both humans and other bots. As digital critical infrastructure, social media needs more effective means to separate human-sourced users from bots and to provide humans the transparency, control, and tools they need to seek trustworthy information.

As AI becomes more powerful, independent, and prevalent, how do we ensure that humans prevail in cyberspace? The changing ethnology of cyberspace demands that we consider this question seriously, as the rise of AI changes both the game and the stakes at all levels. This deserves our attention *now* as in the end, humans are ultimately accountable—we will always be left holding the bag. While writing this book, I asked Google how humans can prevail. Google answered with its own AI:

Fran prompt: How can humans prevail in cyberspace?
Google Search's AI: To prevail in cyberspace, humans must pivot from a technology-first approach to a human-centric one, recognizing that human behavior is both a vulnerability and the key to resilient security. This strategy moves beyond simply preventing attacks to building an ecosystem that can anticipate, withstand, and recover from inevitable breaches while upholding human values.

Even Google's chatbot "agrees" that humans must come first and that change is required to get there.

In a better world, tech is the useful tool we've painstakingly and enthusiastically created, not a taskmaster, predator, overseer, or god. Humans must prioritize themselves, not their increasingly sophisticated cartoon user profiles or AI-driven bots as they design, develop, deploy, and manage technology. We need more humans in cyberspace, and they need more power and control, not less, to promote the interests of themselves and their fellow humans.

We may not have answers to the challenges of cyberspace's changing ethnology today, but *good answers start with serious questions and clear priorities.* We can work to build and manage our essential technologies to transparently identify the bots among us and guide supportive rather than gang behavior. We can give humans more control over outcomes and effective workarounds when our user representations are insufficient. We can focus on getting the best from each resident of cyberspace, adding effective friction or seamlessness as appropriate to increase trust and promote the human experience. We can begin to build a cyberspace where humans come first. And we can get closer to that goal than we are now.

MUSING 2: RETHINKING THE CYBERSPACE ECOSYSTEM

If you squint from 50,000 feet, you can imagine the cascading risks from harmful digital technologies—bias in facial recognition systems, misinformation in social media, exploitation of security vulnerabilities—as invasive cancers that threaten the health of the cyberspace ecosystem. Can we use insights from medical science to reframe how we deal with such risks? It's an intriguing conjecture.

One of my favorite articles is "Cancer's Invasion Equation," a compellingly written *New Yorker* piece from 2017 by well-known oncologist Siddhartha Mukherjee.[6] Mukherjee starts the piece by describing the beautiful and detrimental changes to Lake Michigan in 2011: The water turned crystal clear, so much so that the bottom—replete with sunken ships—was easily visible. Plankton and other organisms that grew in the lake had been consumed by zebra mussels and quagga mussels—two species of mollusks thought to have originally come to the Great Lakes attached to cargo ships from Ukraine.

The mussels were malignant invaders, especially the quagga, changing the ecosystem of Lake Michigan, but interestingly not as invasive in the basins and deltas of Ukraine. Mukherjee makes two key points: First, "the 'invasiveness' of an organism is always a relative concept. . . . An aggressor in one environment is a placid resident in another." Second, when considering how aggressive or harmful a mollusk might be, *the ecosystem itself* is part of the equation: "We want to know which mollusk we're dealing with; but we also need to know which lake."[7]

Mukherjee uses the Lake Michigan analogy to talk about the invasiveness of metastatic cancer. (Stay with me, this actually has something to do with cyberspace.) He cites Stephen Paget, who coined the phrase "seed and soil" to describe the ecosystem in which cancer is the seed and the body is the soil. Many people have abnormal cells associated with cancer—some of which have the potential to grow and divide uncontrollably. Not everyone with these cells develops cancer. In certain bodies, a cancer "seed" flourishes in the body's "soil"; in other bodies cancer is indolent, fails to flourish, and does not progress into a malignancy that overwhelms healthy functioning.

Mukherjee describes three strategies for treating cancer: 1) *attack the seed*—the cancer—and reduce its risk by killing or removing it, retarding its growth, or neutralizing it in some way; 2) *enhance the soil*—the immune system—so it provides an unfavorable environment for the cancer to grow; 3) *get better at predicting the ecological relationship between seed and soil* to determine whether your cancer will flourish or fail to overwhelm healthy cells in your body, key to assessing whether you will ultimately need aggressive treatments like chemotherapy to deal with it.

The seed and soil analogy in cancer is intriguingly relevant to cyberspace, where the severity and spread of invasive technologies with unchecked risks—addiction, misinformation, bullying, data exploitation—can overwhelm the positive aspects of the products and services we use. Just as the ecological relationship between a cancer and its host influences its potential for health risk and negative outcomes, the ecological relationship between an invasive technology and the humans who use it influences its potential for human risk and negative outcomes. If cancer's "invasion equation" depends in part on the way cancer engages with its host, high-risk technologies' "invasion equation" depends on how humans and societies use and manage them in cyberspace.

Why is this analogy relevant? Cancer treatments based on the "seed and soil" model might provide a way to reframe cyberspace and additional strategies to deal with the high-risk, invasive digital critical infrastructure.

What would that look like?

Many cancer treatments attack the seed: Chemotherapy, surgery, radiation and other treatments seek to kill or remove cancer cells, retard their growth, or neutralize them in some way that prevents them from dividing, growing, and taking over. Killing the seed doesn't really work in cyberspace. We're unlikely to "kill" mature technologies by removing them or reducing their popularity (retarding their growth). Once a technology becomes popular (think Facebook, Twitter/X, TikTok), reverse engineering or redesigning it to be fundamentally more private, more secure, less addictive, or more public interest–focused is difficult and likely economically infeasible, as we saw in chapter 3.

We *can* "neutralize" existing invasive technologies (i.e., make them less risky) through responsible use, policy, and regulation, and more user control, education, and awareness. The Knight Institute on behalf of my colleague at UMass Amherst Ethan Zuckerman recently filed a lawsuit against Facebook. The issue? User control of the social media platform. The suit asked the court to recognize that section 230, a part of the Communications Decency Act of 1996, protects the development of tools that empower social media users to control what they see online.

Zuckerman, a rock star public interest technologist, wants to release a tool called Unfollow Everything 2.0, a browser extension that lets users effectively turn off their Facebook newsfeeds, using the platform primarily to connect with their own friends and communities at their own discretion. In effect, the tool would allow users to transform Facebook—an invasive technology with the potential for malignant manipulation—to be more benign, more user-controlled, and less exploitative. The suit was dismissed without prejudice (meaning Zuckerman and the Knight Institute can refile) to give Zuckerman a chance to develop the software and let the court determine whether it should be allowed to interface with Facebook.[8] By the time you read this, the next steps will have played out, but Zuckerman's clever hack to make Facebook more beneficial and less risky constitutes an important neutralization strategy for one of the most invasive malignant seeds of cyberspace.

"Soil-enhancing" strategies create an environment where cancer is more likely to fail. Immunotherapy, in which the patient's own immune system is activated to target cancer cells, is a type of treatment that seeks to alter the ecology of the host to make it a less fertile environment for cancer to thrive.[9] We can alter the ecology of cyberspace by providing lower-risk alternatives for the high-risk, invasive technologies on which we depend. DuckDuckGo is a less privacy-exploiting browser alternative than Google's Chrome, C-SPAN seeks to present unedited public affairs television without commentary, and the e-newsletter *Tangle* provides perspectives from both the right and the left. Both C-SPAN and *Tangle* are less biased than either Fox News or MS NOW (formerly MSNBC). With the transformation of Twitter to X, alternative social media platforms like Bluesky have sprouted up and Reddit has seen steady growth, both providing competitive alternatives for social media and information. More and healthier alternatives give users a choice, and intentional users can choose technologies that lower their risks in cyberspace. A virtuous cycle can be created—the more low-risk alternatives there are to harmful technologies, the better and more competitive they will be.

The third strategy—predicting which seeds will grow in which soils, or who with cancer will actually get sick from it, is much less advanced in the medical ecosystem. This is also true in cyberspace—often risks are not discernable until we gain experience with them at scale and/or "in the wild." Social media is safer when everyone you connect with is someone you know; at scale it's become a fertile environment for predators. It took time to understand the risks of inaccurate facial recognition and the need to manage the use of these systems to reduce bias. The strategies from part II provide "an ounce of prevention" that can help create an environment where technology is less likely to be invasive.

Sometimes things are easier to predict: We can, with a lot of certainty, predict that AI-driven technologies will become more prevalent as digital critical infrastructure. We will need to proactively design, develop, and manage them intentionally to lower human risk. Which products and services will dominate and need the most risk management is unknown, one reason why EU's AI Act (described in chapter 4) calls for the greatest public scrutiny for AI technologies with the most severe risks. The AI Act targets both the seed (the way we design and develop high-risk AI products and services) and the soil (the penalties and layers of vetting required to ensure that risks are managed).

Although we can't predict which AI-driven products and services will become invasive, we can plan for this scenario nonetheless, and in doing so we can reduce the risks of AI-driven digital critical infrastructure. We can adopt a public interest tech mindset and use all the strategies from part II to create and manage AI-driven products and services to promote human well-being. We can devote comparable time and effort to promoting human agency, protections, and the common good as we do to advancing the power and capabilities of AI. Prioritizing the public interest early and often will mean that if/when AI-driven products and services do serve as digital critical infrastructure, they can increase the benefits of cyberspace without dramatically increasing its risks.

Mukherjee's article indicates that oncologists use all the tools at their disposal to combat invasive cancers. We can do the same in cyberspace to make it a healthier habitat and drive a positive ecological relationship between technology, humans, and society. While oncologists generally won't tell you that your cancer is "cured," they will tell you that your treatments were successful in that there is no visible evidence of cancer. When we use multiple strategies to nudge cyberspace toward societal well-being, we are not seeking to cure it or to kill parts of it. We are seeking to treat it with the many tools at our disposal to make it the healthy habitat that we deserve. Keeping the body healthy, whether or not you have cancer, is a persistent practice. So is nudging cyberspace toward the common good.

EVERYONE CAN MAKE A DIFFERENCE

After reading this book, you might be thinking: OK, I buy the idea of social responsibility in cyberspace, but I can't imagine that anything I do will make a difference. I am not powerful, nor do I have tremendous resources. Cyberspace is a tsunami. It's too big and I have no control.

But watching society diminish is not a solution. Cyberspace is as big as the global ecosystem, one that has changed how we operate, our economy, the power dynamics of governments and the private sector, and just about everything else. Seemingly way too big for you and me to impact in any major way, a recognition that sometimes makes positive change feel hopeless. But cyberspace is also as small as every click you and I make, as small as accepting or rejecting cookies on a website, vetting something we read on X, view on TikTok, or learn from ChatGPT, as small as adding two-factor authentication to protect an online bank or email account. A quote often attributed to the anthropologist Margaret Mead is, "Never doubt that a small group of thoughtful, committed citizens can change the world. Indeed, it is the only thing that ever has." Although the focus of the quote is cultural change, it also applies to change in the cyberspace ecosystem. And the more of us who intentionally engage with technology responsibly, the bigger and more positive our impact will be.

If enough of us adopt a public interest tech mindset, find ways to lower human risk, make responsible choices, and work toward promoting the common good in what we do, our efforts will mount up; our cumulative hivemind will begin to approach things in a more beneficial way. More will join in. Our actions will make a positive difference.

Now is the time for us to rally around better tech—a cyberspace that promotes the well-being of the humans within it. I invite us—the human creators and ultimate beneficiaries (and victims) of cyberspace—to reframe how we approach our digital present and future, to reconsider who cyberspace products and services should benefit, and to reprioritize people as we create, manage, and use the digital technologies essential to societal advancement. All is not lost. In a real sense, it is up to us to reduce risks and promote human benefits, to not get discouraged when our efforts seem too small or not enough, to stay consistent, and to give it our best shot. Everyone can do something. The first step to real change is believing you can do it.

ACKNOWLEDGMENTS

This book was born at the Radcliffe Institute and raised at UMass Amherst and more coffee shops than I can count. Its midwives and godmothers were the VPs—Liz Chiarello, Alexi Lahav, and Anne Higonnet—dear friends, fellow Radcliffe fellows, and excellent writers all. Wonderful colleagues and friends participated in the book's upbringing, discussing ideas, reading drafts of the text, and offering wisdom and comments. Heartfelt thanks to Josh Greenberg, Thatcher Drew, Jim Kurose, Emily Nutwell, Michelle Ciccone, Donna Baron, and Immaculata De Vivo for their generous feedback and support. Much gratitude as well to Bruce Schneier, Laura Quilter, Susan Rabiner, and Lee Kravetz for excellent book parenting advice. Many years ago, Hilary Mason told me that the best way to get my ideas into the public discourse was to write a book. Thanks Hilary for planting the seed.

Like many beloved children, raising this book took a village. Thanks to Kathy Pham, David Jensen, Frank Pasquale, Kiel Brennan-Marquez, Emily Nutwell, and Ethan Zuckerman for their inputs to and appearances in the book. Special thanks to the "PIT Crew" at UMass—Donna Baron, Carolina Rossini, Charlie Schweik, Ethan Zuckerman, Emily Nutwell, and Michelle Ciccone as well as Laura Haas, Karl Rethemeyer, Anne Massey, Keith Marzullo, Kumble Subbaswamy, and the UMass administration, faculty, and staff for believing in us and supporting us.

Thanks to all of the students in the UMass "Introduction to PIT" course, PIT clubs, and the PIT New England Summer Institute for sharing their passion for a better world—you are the future and give me hope. Likewise, many thanks to my great community of public interest tech–focused colleagues and inspirations—the Berkman Klein Center hivemind and Public Interest Technology Working Group, PIT New England colleagues, Radcliffe Fellows class of 2019–2020 and summer 2023, colleagues in PIT-UN, and especially Azer Bestavros, Ziba Cranmer, Michelle Shevin, Anne Washington, Colette Basiliere, Jill Lepore, Michelle Trim, Erhardt Graeff, Rebecca Tabasky, Danny Goroff, Andreen Soley, Jenny Toomey, Brian Schottlaender, Alyssa Goodman, Chris Borgman, and many more.

Writing a book is so much more than putting thoughts to words and words on a page. To the fabulous folks at MIT Press and especially to Amy Brand, Gita Devi Manaktala, Ginny Crossman, Suraiya Jetha, Janice Audet, Cassidy Sulaiman, David Luljak, and many others, thank you for helping me navigate my book publishing journey and ensuring that this book survived adolescence to adulthood. Neither of us could have done this without you. Hopefully the book is now eligible to vote.

Public interest technologists are raised, not born. I am incredibly grateful to so many people over the course of my career—colleagues at UCSD and the San Diego Supercomputer Center, at Harvard, RPI, and UMass, at the Sloan Foundation, the Research Data Alliance, and the Library of Congress, NEH, and NSF for being role models and supporters. Thank you for the work you do every day—work that sets the standard for how to use a public interest tech mindset to make science and society better. I am also grateful to you, the readers. Thanks for your interest in responsible tech and for reading this book. I hope it provides ideas for you to make an even bigger and more positive impact in your work and lives.

Finally, I dedicate this book with my whole heart to Emily, Nick, and Ekta, my personal advisory committee, my inspirations, and my own source of hope for the future. Thank you for all you do, your thoughtfulness, your wisdom, your integrity, and all your love and support. You make my world a better place.

APPENDIX: READER'S GUIDE / TEACHING RESOURCE

The goal of this guide and teaching resource is to help readers and students develop a "public interest tech mindset"—a way of thinking about the ever-changing world we live in, and a structured way to assess and react to the news of the moment and the tech that influences our lives.

Tech advances quickly and it's sometimes hard to know what is hype and what is real. In my courses, I ask students to keep a journal and respond to prompts that help them develop critical thinking skills about the sociotechnical world. Both book groups and academic classes can use prompts like these to seed discussion about tech that is new, relevant, impactful, or important in the moment as well as in the long run.

Journal prompts that have been useful and relevant to the themes in this book for my classes include the ones below. I encourage book groups and instructors to use these as a point of departure, and to create their own prompts that spur discussion and a deeper understanding of the potential opportunities, risks, and societal impacts of the tech they are dealing with in their work and lives. In my own classes, students are not allowed to use AI assistants to respond to journal prompts as I am particularly interested in their own thoughts and experiences.

(They are allowed to use AI for other kinds of assignments as described later.) Many thanks to the insightful Michelle Ciccone and Emily Nutwell for their partnership in creating the course materials and curricula below.

PROMPTS FOR COURSES AND BOOK GROUPS

GENERAL PROMPTS

- Pick your favorite tech service or product.
  - How do you and others use that technology?
  - What have been its positive societal consequences?
  - What have been its negative societal consequences?
  - Are there consequences that were likely unanticipated when the product/service was launched?

- Is there a tech tool or service that you wish you had that is not commercially available? What would it do and why would it be a benefit?

- What excites you most and what worries you most about tech?

- Pick a new information technology and reflect on the following:
  - What is its technological innovation?
  - What choices were likely made in its design and sourcing?
  - What public interests does it promote; what public interests does it put at risk?
  - Is the new technology more beneficial for some groups than others?

- What systems (specifically) do you use that are biased and how does that bias manifest itself?
  - How does this bias impact you?
  - What are the societal risks of this bias?
  - How do you mitigate the risks of this bias?

- How do you stay informed about new developments in technology? How do you determine what is accurate and trustworthy and what is not? What do you do to find trustworthy information?

- How important is it for the general public to know about information technology? What skills and information should they have and where should they get them? What is the responsibility of academia, the private sector, and the public sector in ensuring that the general public is tech-literate?

AI PROMPTS

- What excites you and what scares you the most about AI?
- With respect to AI, what do you think is hype and what do you think is real? How are you staying informed about developments in AI? How do you make sure your information sources are trustworthy?
- Do you think that AI will eventually achieve superintelligence? Human intelligence? How did you arrive at your opinion?
- How should your organization/university/group develop policy for AI use? What key elements would you include in the policy and what do you think would be over-restrictive?
- How should governments deal with today's AI innovations? What actions should be taken to protect the public? What would the impact of these actions be on the private sector?

SOCIAL MEDIA PROMPTS

- How do you use social media?
    - What have been the benefits for you and for society in using social media?
    - What have been the downsides for you and for society in using social media?
- What would you do to improve the integrity of content on your favorite social media platform?
- Should social media (Facebook, TikTok, Reddit, Bluesky, X, Truth Social, etc.) be managed as a publisher (accountable for the integrity of content)

or as a platform (a pass-through, not accountable for the integrity of content)? What would be the consequences of going in either direction?

SEARCH PROMPTS

- Do a Google Search of yourself. What appears in that Google Search?
  - Does what appears in your Google Search match what you would expect? Anything unexpected or concerning?
  - What's missing? What's incorrect?
  - What impression might someone get about you based on these Google Search results?
  - Do the same thing with a different web browser and/or a different search engine. Is there anything different about what appears in the results?
- Pair up with a friend and do a Google Search of the same topic and compare your search results.
  - Did you get the same results? What might account for any differences?
  - What are the pros and cons of Google Search results being tailored to each user?
- What have been the positive and negative consequences of search engines? What have we gained as individuals and society and what have we lost as these tools have become indispensable to the pursuit of knowledge? How is AI changing the way we search? Does AI make search better?

DATA AND PRIVACY PROMPTS

- Go to myadcenter.google.com. Click on "Manage Privacy" from the left-hand menu. Scroll down the page to see "Categories used to show you ads." (Note: If you have multiple Google accounts [e.g., personal and school or work], look at this page from each account to see if there are any differences.) Reflect:
  - Are these categories correct?
  - If some are incorrect, why might Google categorize you in these ways?

  - Do these categories match what appears in your Google Search results? What might account for differences?
  - Does what you found impact the way you think about Google's tracking?

- What do you do now to manage your data profile?
  - What strategies might you add to what you are already doing to promote more personal privacy?
  - In a perfect world, what kind of controls do you wish you could have? How would having those controls change things for users, the private sector, and in the public sector?

- Consider the European Union's General Data Protection Regulation (GDPR).
  - Describe a right or privilege you would have under the GDPR in the EU that you do not have in your home country (or in another country if you are from the EU). Why would that be an important digital right for you to have?
  - Are there digital rights that you have or would like to have that EU citizens do not have?

- Reflect on your thoughts about privacy:
  - How much does privacy matter to you?
  - What about you should be private (not shared without your permission)?
  - What are two societal interventions that might be deployed to help promote your privacy? What are potential consequences if those interventions were deployed?

- "Bystander surveillance" (capturing the images of people in the course of broadcasting something else—think webcams) in public places is not illegal but provides information that may be analyzed, shared, or exchanged.
  - Give an example of bystander surveillance that involves you.
  - What, if any, instances of bystander surveillance do you think are inappropriate?

- ◦ What societal controls might be developed to protect individuals from inappropriate bystander surveillance or bystander surveillance that puts them at risk?

PIT AND THE WORKFORCE PROMPTS

- How does AI impact your (current or prospective) job? What, if anything, will you need to know about AI to do the work you want to do? What are the ethical and safety issues associated with using AI for your job?
- Write about an exemplary public interest technologist that you know or have heard of. How do they use a public interest tech mindset in their work or lives? What strategies can you learn from their approach?
- Which tech goods, services, or companies do you think do a good job of promoting the public interest? Why? What strategies do they use to promote the public interest?
- Which jobs do you think AI is best equipped to do and which jobs are humans best equipped to do? For jobs that we rely on AI to do, how should we ensure that it is being used safely, ethically, and beneficially?

A GENERAL EDUCATION PUBLIC INTEREST TECH SYLLABUS

My university classes are general education courses designed to help students develop a public interest tech mindset and provide a "tasting menu" of the responsible tech challenges and opportunities from the world they live in. The curriculum is targeted to all undergrads in any major with any level of tech experience; it is not targeted solely to computer scientists. Everyone can develop a public interest tech mindset, and the finance, theater, biology, math, sociology, and other majors who have taken the course have valued learning more about tech and its impacts on the worlds they must navigate. Computer science majors in the course have enjoyed seeing the tech they learn about in more mainstream computer science courses "in the wild" (i.e., embedded in the real world), and have appreciated the opportunity to gain an awareness of its larger societal impacts. Students in the course are also exposed to a variety of guest lecturers from different disciplines and with different perspectives. We save time to "interview" each guest speaker after they present to learn about how they apply a public interest tech mindset to their own work.

The Intro to Public Interest Technology course is customized every time it is given to reflect advances in technology and current tech opportunities and challenges. Although the content of the course varies each time it is taught, the structure is generally the same. A suggested syllabus for the course, based on the course taught at UMass Amherst (CICS 127) is below.

Suggested Syllabus

PART 1—INTRODUCTION TO PUBLIC INTEREST TECHNOLOGY

- What Is the Public? What Is Public Interest Technology?

  The lecture introduces public interest technology and explores how the notion of "public" varies across environments. A range of public interests is described, including fairness, safety, security, privacy, sustainability, ethical treatment, etc.

- Why PIT Matters

  The lecture focuses on how tech is not neutral but the result of many decisions made to advance personal, organizational, or business goals. Lecture examples focus on the societal impacts of these decisions (e.g., Facebook's personalized Feed and Google's ranking algorithms).

- Everything Has Consequences—Societal Impacts of Technology

  Lectures provide specific examples of new technological innovations and the opportunities and challenges they create for society (self-driving cars and the invention of freon are often used as examples). We describe how some tech serves as digital critical infrastructure and talk about our different expectations for commercial products and critical infrastructure.

- Tutorial: Writing and AI

  This lecture helps students learn to use AI to enhance, rather than replace, development of their storytelling abilities, presentation skills, and writer's voice. (More details on this lecture are in the last section of this appendix.)

PART 2—TECHNOLOGY BASICS AND SOCIETAL IMPACTS

- Data and Surveillance

  The lecture(s) focus on how digital data is collected, interpreted, and used. Topics include correlation versus causation, data privacy, bias in facial recognition, and surveillance. Lectures are often accompanied by a "My Data Profile" class exercise.

- AI Basics and AI Safety

  The lecture(s) begin by discussing what AI is and its various subfields. Topics include narrow AI versus general AI; basics of machine learning and neural nets; what AI can, can't, and shouldn't be used for; and a discussion of the challenges of AI governance.

- The Internet and Net Neutrality

  The lecture focuses on the basic structure and governance of the internet and how it forms the foundation of cyberspace. Topics include the societal implications of net neutrality and pay-to-play policies.

- Tech Ethics

  The lecture begins by exploring what, which, and whose ethics should apply to digital technologies. Example ethical dilemmas include utilitarian ethics and the trolley problem, the Google and Project Maven controversy, and a discussion of the Moral Machine experiment.

PART 3—SOCIETAL INTERVENTIONS FOR MANAGING TECH

- Tech Law and Policy

  The lecture focuses on tech law and policy. Differences between law, policy, and regulation are discussed, as is the importance of enforcement. AI law may be used as an example. Topics include the EU AI Act, US AI laws and policies, the OECD AI Principles, and AI policy and law in different jurisdictions around the world.

- Privacy Law in the US, EU, and China

  The lecture focuses on privacy law and policy. The GDPR is described in some detail, as are US federal and state privacy laws and the PIPL in China.

- PIT and the Office of Technology Assessment

  The lecture focuses on the US Office of Technology Assessment and its function before being defunded. The OTA is used as an example of the need for expertise in developing effective law and policy and the dynamic environment in which law, regulation, and policy are created.

- Your Digital Rights

  The lecture focuses on the broader notion of digital rights beyond privacy and AI. The lecture explores how digital rights differ in different jurisdictions around the world and highlights the societal and cultural implications of these differences.

PART 4—PIT WORK AND WORKFORCE

- PIT in the Workforce—Jobs, Trends, Role Models

  The lecture focuses on tech-related jobs and job trends. We explore how a public interest tech mindset can be used in any line of work to promote responsible development, use, and management of tech. Examples of how specific individuals incorporate this mindset into their work are given.

- AI and the Future of Work

  The societal impact of AI-driven work is discussed. Topics include AI's impact on jobs and the workforce, as well as the responsibility and liability challenges of AI-driven work.

- Student PIT Projects

  We often use the end of the semester for students to report on their group projects. This provides an additional setting to practice presentation skills.

CLASS ASSIGNMENTS

The syllabus above is targeted to all undergrads; students are not required to write code for their grades. Instead, we seek to help students develop critical thinking skills about tech and better their communication skills, particularly their presentation and writing skills. In the course, students are graded on both content *and* writing/presenting style, with considerable feedback and guidance on both. In particular, we provide guidance on how to create an effective evidence-based narrative for their topic and what makes a compelling presentation/written piece. In the course, students are asked to do two presentations, to do two to three writing assignments (one of which may be associated with a group project), and to write in their journals weekly. Interested instructors should of course vary the length and specifics of the assignments to fit their own courses and teaching styles.

Example Assignments

- INDIVIDUAL WRITING ASSIGNMENT: Different presidential administrations may take different approaches to tech governance. **Identify a tech-driven or tech-focused area** where the approach taken by the current administration is different from the approach taken by the previous administration. **Compare the current administration's approach with the previous administration's approach.**
    - What are the impacts and consequences of this change in federal approach from a public interest tech perspective?
    - How might citizens be involved in promoting the public interest with respect to this change in approach?
    - Cite articles from reputable, mainstream news sources to support your views in the piece.
    - Students are allowed to use AI to enhance their writing with appropriate documentation (more about this in the next section).
- INDIVIDUAL WRITING ASSIGNMENT: **Describe an area of your life or work where you would like increased privacy.** What strategies (policy,

practice, law, product design, other?) could be developed to ensure that you are better protected? Your piece should:
  - Give examples of what increased privacy would look like and how you would benefit.
  - Describe the public/private interest trade-offs of enforcing your strategies.
  - Include citations for at least two credible outside sources (not AI chatbots) that support your arguments.
  - Students are allowed to use AI to enhance their writing with appropriate documentation.

- INDIVIDUAL WRITING ASSIGNMENT: **Write a personal essay** describing the following:
  - What job do you see yourself doing in five years?
  - Identify three ways that you could promote the public interest through the responsible use of technology in your job.
  - What do you see as the opportunities, challenges, and potential impacts of your efforts?
  - Students are allowed to use AI to enhance their writing with appropriate documentation.

- INDIVIDUAL WRITING ASSIGNMENT: In this class, we have discussed a variety of public interest tech challenges. Pick a specific public interest tech challenge (from class or outside of what we have discussed) that is especially interesting, meaningful, or worrisome to you. **Write a piece** that incorporates the following components/issues:
  - Describe the challenge. What do you see as its risks, benefits, and existing societal impacts?
  - What happens if we don't address this challenge in a way that minimizes the risks to the public?
  - What are specific and actionable societal controls that could be used to address this challenge? How would we implement them?
  - Students are allowed to use AI to enhance their writing with appropriate documentation.

- INDIVIDUAL WRITING ASSIGNMENT: Imagine that you are a guest lecturer in this class. You will be talking about a tech-related topic that is really important to college students and not already covered this semester. **Write a piece** describing your lecture that incorporates the following components:
    - What would you talk about?
    - What information would you include in your lecture? Cite reputable mainstream sources (do not cite AI chatbots).
    - What are the main points and takeaways for students from your lecture?
    - What journal prompt would you give to encourage students to think more deeply about this lecture and personalize its takeaways? What reading might you give to students to accompany this lecture?
    - Students are allowed to use AI to enhance their writing with appropriate documentation.

- INDIVIDUAL WRITING ASSIGNMENT or GROUP WRITING/PRESENTATION ASSIGNMENT: Imagine you (or your group) staffs a Senator or Congressperson (or works in the [now defunded] Office of Technology Assessment). **Select a public interest technology issue** of interest to your stakeholder and **create an evidence-based briefing** on the issue (part 1) and **present your results** in class (part 2).
    - Part 1: Briefing
        - The briefing should include:
            - A crisp explanation of the public interest technology issue.
            - Analysis and greater detail about its impact on society, the economy, the environment, and/or other relevant areas.
            - A description of the pros and cons of various policy approaches
        - The briefing should include citations from credible sources (do not cite AI chatbots).
        - Students are allowed to use AI (with documentation) to enhance development of the briefing.

  - Part 2: Presentation
    - Prepare a short presentation of your briefing for the class (your stakeholder). The presentation should crisply and compellingly describe the issue, key specifics from your analysis, and potential policy approaches.
    - Students are allowed to use AI (with documentation) to enhance development of the presentation.

- GROUP PROJECT: **Create a three to four minute video** featuring a person or organization that makes, manages, or uses technology to promote positive societal impact. (This must be a person or an organization doing something *proactive* to promote the public interest via their approach to technology.)
  - This project is graded on both the video and a written reflection. For the video, all students receive the same grade; the reflections are written by each student and grades are given individually.
  - To help promote success, students are given guidance about effective collaboration and working in teams. They are asked to submit their ideas for interviewees for approval from the instructor prior to starting the project—this ensures that the videos are correctly focused on individuals and organizations that achieve project goals.
  - In the written reflection, students are asked to talk about their personal experience collaborating to develop the video and to describe their role in the collaboration.
  - Students are allowed to use AI (with documentation) to enhance their development of the video and reflection parts of the project.

- IN-CLASS GROUP PROJECT ("PIT ACADEMY AWARDS"): Divide the class into groups of four. Each group is in the "Academy of Public Interest Technology" (APIT). Give the groups prep time to:
  - Create a category for public interest technologists, organizations, services, or products for the APIT Awards (e.g., "Best PIT Book," "Best PIT Product," "Best PIT For-Profit Company").

- ◦ Identify three great nominees for the chosen category and prepare slides with the nominees and a short description of why each of them should win.
- ◦ When prep time is over, each group presents their category and nominees, and the class votes on the winners.

STUDENTS AND AI

In my classes, students are not permitted to use AI assistants for their journal writing. Journal prompts are designed to focus on students' personal experience and thoughts about tech, something that AI is less likely to be useful for. Students *are* permitted to use AI assistants on other writing assignments. I want to help them become both better writers and more proficient at using AI tools. These tools are everywhere and are used by an increasing number of professionals. They can be useful for getting to "good enough" for writing, code, and ideas, but it is important for the student to develop their own skills to achieve "better" (and, importantly, to be able to knowledgeably evaluate the responses they get from chatbots). AI is not always accurate, and it is certainly not unbiased. My goal in the classroom is to help students use AI as a tool to *enhance, but not replace,* their own conceptual and writing skills, which I'd like to help them improve.

I seek to accomplish this in two ways: First, I allow the use of AI on longer non-journal written assignments and grade these assignments on both the content and the quality of their writing (and the development of their "voice"). Second, I do a "writing with AI" tutorial lecture (before they are given their writing assignments) that describes the benefits and liabilities of using AI and shows them (with examples) how they might use AI in their own writing. Key points of the AI tutorial lecture include:

- Spend some time thinking about the assignment and *start with a draft of your own ideas.* AI should *not* be used to conceptualize and write the whole piece—that doesn't expand your conceptual skills or develop your writing voice, and AI writing can be too homogenized to compel the reader.

- AI can be useful as a *developmental editor* for your writing, as a *generator of ideas* to supplement your own conceptualization, or to *create a more cohesive draft* of a piece (very useful for groups of students with different voices cowriting a piece or for students for whom English is not their primary language).
  - To use AI as a **developmental editor**, students can provide text and prompts on language structure, style, or content—"fit this into a word limit," "make this writing more compelling," or "what is missing?"—and use the response to enhance the current draft of the piece. The AI responses should always be vetted and used as a suggestion, not the final product.
  - To use AI as an **idea generator**, students should start with their own ideas and supplement their efforts with AI. Students can give prompts like "give three examples of data broker laws" or "how does facial recognition bias impact communities?" The responses should be vetted and used as a suggestion, and students should ask for citations and verify/follow up on them. Starting with AI rather than students' own ideas doesn't help them build the critical thinking or conceptual skills needed for professional work, nor does it help them gain the knowledge needed to effectively evaluate the outputs of AI.
  - To use AI to **create a more cohesive draft** of a piece, prompts such as "harmonize the following text so it has a single voice" or "reduce the number of lists in this paragraph" can be used. As before, the results should be checked to ensure that meaning has not been lost or mangled, and AI outputs should be integrated into the piece so that the resulting narrative has a unified and compelling voice. (With group writing, choosing a single editor from the group can also accomplish this goal.)

- The privacy policies of many AI tools and services allow providers to collect your prompts and related data. This information may be used to feed learning models, shared with service providers or partners, and/or supplemented via tracking and cookies. As a matter of AI safety, you should never include sensitive personal information (e.g., your social security number or personal documents) as part of a prompt. You may or may not be able to opt out of a company's use of your information and *you should only use an AI tool if you are comfortable with its privacy policy.*

- As Bender and colleagues state, AI is a "stochastic parrot."[1] It does not "know" what it is giving you. The user must assess the veracity, appropriateness, and usefulness of the outputs the AI tool provides.
- AI can plagiarize from what it has learned and make up material and sources that don't exist. The user must check and verify *all* information that they incorporate into their work ("*Don't trust* and do verify.").

During the AI tutorial lecture, I provide specific examples of how AI can be used as an idea generator, as a developmental editor, or to improve the cohesiveness of a piece. In my examples, I sometimes include a particularly interesting response I got to a demonstration prompt: "rewrite the first paragraph of my academic bio in the voice of Taylor Swift." I got a song that wouldn't pass muster with anyone . . . .

After this lecture, students have some basis for using AI to enhance their own writing. As part of their writing assignments, students can choose whether or not they want to use AI. If they use AI, *they must include documentation of all prompts given to AI as an appendix* to their work so we can check to see how they are using the tool.

Incorporating AI in this way in the course seems to provide a kind of "safe space" for students to improve their writing and experiment with AI. I have been surprised that many students choose not to use AI tools, taking the opportunity to improve their own skills without the potential of a crutch. Sometimes, we suggest students use AI as part of our feedback when we think it would help. Sometimes students have used it and it is clear their assignment has too much "AI writing" when their own voice is more interesting. We provide that feedback too.

AI is not omniscient and neither is it always better than you are. It is a powerful tool for all of us and should be used in a way that emphasizes its benefits to us. I ask students to think of AI like a chainsaw. You can't build a whole house with a chainsaw, but for cutting large amounts or big pieces of wood, it's an incredible help. (Don't try to use it for plumbing or electricity . . .) I try to help the students develop their own skills as well as gain a more sophisticated sense of when and where AI can add value: when it *can* be used, when it *can't* be used, and when it *shouldn't* be used—basics of AI literacy and safety they should know to get the best from this important tool in school and beyond.

NOTES

Prologue, Part I

1. Merrit Kennedy, "Lead-Laced Water in Flint: A Step-by-Step Look at the Makings of a Crisis," *NPR*, April 20, 2016, https://www.npr.org/sections/thetwo-way/2016/04/20/465545378/lead-laced-water-in-flint-a-step-by-step-look-at-the-makings-of-a-crisis.

Chapter 1

1. Wikipedia, "Cultural Lag," last modified November 22, 2024, 15:22 (UTC), https://en.wikipedia.org/wiki/Cultural_lag.

2. *Britannica*, "Tim Berners-Lee," last updated July 22, 2025, https://www.britannica.com/biography/Tim-Berners-Lee.

3. Martin Hilbert and Priscila López, "The World's Technological Capacity to Store, Communicate, and Compute Information," *Science* 332, no. 6025 (2011): 60–65, https://doi.org/10.1126/science.1200970.

4. William Gibson, *Neuromancer* (New York: Ace Books, 1984).

5. Joint Chiefs of Staff, *Cyberspace Operations*, Joint Publication 3–12 (R) (Washington, DC: Joint Chiefs of Staff, 2013), https://irp.fas.org/doddir/dod/jp3_12r.pdf.

6. David McCabe, "Google Search Results Vary Widely by User, New Study Says," *Axios*, December 4, 2018, https://www.axios.com/2018/12/04/google-search-results-vary-widely-by-user-new-study-says.

7. Justin Sherman, "How Shady Companies Guess Your Religion, Sexual Orientation, and Mental Health," *Slate*, April 26, 2023, https://slate.com/technology/2023/04/data-broker-inference-privacy-legislation.html.

8. Anita Ramasastry, "Websites That Charge Different Customers Different Prices: Is Their 'Price Customization' Illegal? Should It Be?," *FindLaw* (blog), June 20, 2005, https://supreme.findlaw.com/legal-commentary/websites-that-charge-different-customers-different-prices.html.

9. Federal Trade Commission, "Behind the FTC's Inquiry into Surveillance Pricing Practices," *FTC Office of Technology Blog*, July 23, 2024, https://www.ftc.gov/policy/advocacy-research/tech-at-ftc/2024/07/behind-ftcs-inquiry-surveillance-pricing-practices.

10. Jonathan Rothwell, "Teens Spend Average of 4.8 Hours on Social Media Per Day," *Gallup News*, October 13, 2023, https://news.gallup.com/poll/512576/teens-spend-average-hours-social-media-per-day.aspx.

11. Emily A. Vogels, Risa Gelles-Watnick, and Navid Massarat, *Teens, Social Media and Technology 2022*, Pew Research Center, August 10, 2022, https://www.pewresearch.org/internet/2022/08/10/teens-social-media-and-technology-2022/.

12. Freya India and Jon Haidt, "Algorithms Hijacked My Generation. I Fear for Gen Alpha," *After Babel* (blog), November 10, 2023, https://www.afterbabel.com/p/algorithms-hijacked-my-generation.

13. Federal Bureau of Investigation, "Sextortion: A Growing Threat Preying Upon Our Nation's Teens," press release, January 17, 2024, https://www.fbi.gov/contact-us/field-offices/sacramento/news/sextortion-a-growing-threat-preying-upon-our-nations-teens.

14. Olivia Carville, "Scammers Are Targeting Teenage Boys on Social Media—and Driving Some to Suicide," *Bloomberg*, April 15, 2024, https://www.bloomberg.com/features/2024-sextortion-teen-suicides/.

15. Emma Henderson Vaughan, "NCMEC Releases New Sextortion Data," *National Center for Missing and Exploited Children* (blog), April 15, 2024, https://www.missingkids.org/blog/2024/ncmec-releases-new-sextortion-data.

16. Cecilia Kang and Natasha Singer, "Meta Accused by States of Using Features to Lure Children to Instagram and Facebook," *New York Times*, October 24, 2023, https://www.nytimes.com/2023/10/24/technology/states-lawsuit-children-instagram-facebook.html.

17. Natasha Singer, "How Mark Zuckerberg's Meta Failed Children on Safety, States Say," *New York Times*, June 22, 2024, https://www.nytimes.com/2024/06/22/technology/zuckerberg-instagram-child-safety-lawsuits.html.

18. New Mexico Department of Justice, "New Mexico Attorney General Raúl Torrez Announces Arrests of Suspected Online Predators Following 'Operation MetaPhile,'" press release, May 8, 2024, https://nmdoj.gov/press-release/new-mexico-attorney-general-raul-torrez-announces-arrests-of-suspected-online-predators-following-operation-metaphile/.

19. Singer, "How Mark Zuckerberg's Meta Failed Children on Safety."

20. Sarah Miller, "The Addictiveness of Social Media: How Teens Get Hooked," Jefferson Health, June 2, 2022, https://www.jeffersonhealth.org/your-health/living-well/the-addictiveness-of-social-media-how-teens-get-hooked; Trevor Haynes, "Dopamine, Smartphones & You: A Battle for Your Time," Harvard University, *Science in the News* (blog), May 1, 2018, https://sites.harvard.edu/sitn/2018/05/01/dopamine-smartphones-battle-time/.

21. Isobel Asher Hamilton, "Silicon Valley Insiders Say Facebook, Snapchat, and Twitter Are Using 'Behavioral Cocaine' to Turn People into Addicts," *Business Insider*, July 4, 2018, https://www.businessinsider.com/silicon-valley-insiders-tell-bbc-how-tech-firms-turn-users-into-addicts-2018-7.

22. Ronald E. Dahl, Nicholas B. Allen, Linda Wilbrecht, and Ahna Ballonoff Suleiman, "Importance of Investing in Adolescence from a Developmental Science Perspective," *Nature* 554 (2018): 441–450, https://doi.org/10.1038/nature25770.

23. Emily A. Vogels, *The State of Online Harassment*, Pew Research Center, January 2021, https://www.pewresearch.org/internet/wp-content/uploads/sites/9/2021/01/PI_2021.01.13_Online-Harassment_FINAL-1.pdf.

24. Andrew P. Smith and Hasah Alheneidi, "The Internet and Loneliness," *AMA Journal of Ethics* 25, no. 11 (2023): 833–838, https://doi.org/10.1001/amajethics.2023.833.

25. Future of Life Institute, "Pause Giant AI Experiments: An Open Letter," March 22, 2023, https://futureoflife.org/open-letter/pause-giant-ai-experiments/.

Chapter 2

1. Adam Clark Estes, "Why the Internet (Probably) Won't Break During the Coronavirus Pandemic," *Vox*, March 25, 2020, https://www.vox.com/recode/2020/3/25/21188391/internet-surge-traffic-coronavirus-pandemic.

2. Shannon Bond, "A Must for Millions, Zoom Has a Dark Side—and an FBI Warning," *NPR*, April 3, 2020, https://www.npr.org/2020/04/03/826129520/a-must-for-millions-zoom-has-a-dark-side-and-an-fbi-warning.

3. Cybersecurity and Infrastructure Security Agency, "Critical Infrastructure Sectors," accessed September 4, 2025, https://www.cisa.gov/topics/critical-infrastructure-security-and-resilience/critical-infrastructure-sectors.

4. Cybersecurity and Infrastructure Security Agency, "Critical Infrastructure Sectors."

5. Brian Fung, "We Finally Know What Caused the Global Tech Outage—and How Much It Cost," *CNN*, July 24, 2024, https://www.cnn.com/2024/07/24/tech/crowdstrike-outage-cost-cause/index.html.

6. US Department of Agriculture, "Recalls and Public Health Alerts," accessed September 4, 2025, https://www.fsis.usda.gov/recalls.

7. Alex Hern, "Hacking Risk Leads to Recall of 500,000 Pacemakers Due to Patient Death Fears," *The Guardian*, August 31, 2017, https://www.theguardian.com/technology/2017/aug/31/hacking-risk-recall-pacemakers-patient-death-fears-fda-firmware-update.

8. Eduard Kovacs, "Security Firm Discloses Medical Device Flaws as Part of Investment Strategy," *SecurityWeek*, August 26, 2016, https://www.securityweek.com/security-firm-discloses-medical-device-flaws-part-investment-strategy/.

9. BioSpace, "FDA Recalls 465,000 St. Jude Medical Pacemakers over Hacking Fears," press release, August 31, 2017, https://www.biospace.com/fda-recalls-465-000-st-jude-medical-pacemakers-over-hacking-fears.

10. Dave Fornell, "FDA Confirms Cybersecurity Vulnerabilities of St. Jude's Implantable Cardiac Devices, Merlin Transmitter," *Diagnostic and Interventional Cardiology*, January 9, 2017, https://www.dicardiology.com/article/fda-confirms-cybersecurity-vulnerabilities-st-judes-implantable-cardiac-devices-merlin.

11. Brian P. Dunleavy, "Pacemakers, Insulin Pumps Can Be Hacked, Experts Say," *United Press International*, June 1, 2022, https://www.upi.com/Health_News/2022/06/01/medical-devices-pacemakers-cybersecurity/7041653656330/.

12. Dunleavy, "Pacemakers, Insulin Pumps Can Be Hacked"; Cybersecurity and Infrastructure Security Agency, "Cyber Threats to Medical Technology and Communication Technology Protocols," infographic, CISA, February 2024, https://www.cisa.gov/sites/default/files/2024-02/Cyber%20Threats%20to%20Medical%20Technology%20and%20Communication%20Technology%20Protocols%20%28508%29.pdf.

13. Stephen Neely, Christina Eldredge, and Ron Sanders, "Health Information Seeking Behaviors on Social Media During the COVID-19 Pandemic Among American Social Networking Site Users: Survey Study," *Journal of Medical Internet Research* 23, no. 6 (2021), https://doi.org/10.2196/29802.

14. Ed Pertwee, Clarissa Simas, and Heidi J. Larson, "An Epidemic of Uncertainty: Rumors, Conspiracy Theories and Vaccine Hesitancy," *Nature Medicine* 28, no. 3 (2022): 456–459, https://doi.org/10.1038/s41591-022-01728-z.

15. Katie Rogers, Christine Hauser, Alan Yuhas, and Maggie Haberman, "Trump's Suggestion That Disinfectants Could Be Used to Treat Coronavirus Prompts Aggressive Pushback," *New York Times*, April 24, 2020, https://www.nytimes.com/2020/04/24/us/politics/trump-inject-disinfectant-bleach-coronavirus.html.

16. Neely, Eldredge, and Sanders, "Health Information Seeking Behaviors."

17. Kalhan Rosenblatt, "ChatGPT Passes MBA Exam Given by a Wharton Professor," *NBC News*, January 23, 2023, https://www.nbcnews.com/tech/tech-news/chatgpt-passes-mba-exam-wharton-professor-rcna67036.

18. Kevin Roose, "A Conversation with Bing's Chatbot Left Me Deeply Unsettled," *New York Times*, February 16, 2023, https://www.nytimes.com/2023/02/16/technology/bing-chatbot-microsoft-chatgpt.html.

19. Molly Bohannon, "Lawyer Used ChatGPT in Court—and Cited Fake Cases. A Judge Is Considering Sanctions," *Forbes*, June 8, 2023, https://www.forbes.com/sites/mollybohannon/2023/06/08/lawyer-used-chatgpt-in-court-and-cited-fake-cases-a-judge-is-considering-sanctions/.

20. European Parliament, "EU AI Act: First Regulation on Artificial Intelligence," Topics, last updated February 19, 2025, https://www.europarl.europa.eu/topics/en/article/20230601STO93804/eu-ai-act-first-regulation-on-artificial-intelligence.

21. "Data: Coronavirus and School Closures in 2019–2020," *Education Week*, January 10, 2022, https://www.edweek.org/data-coronavirus-and-school-closures-in-2019-2020/2021/12.

22. Michael Calabrese and Amir Nasr, "Pandemic Disruption: From Homework Gap to Remote Learning Chasm," in *The Online Learning Equity Gap* (Washington, DC: New America, 2020), https://www.newamerica.org/oti/reports/online-learning-equity-gap/i-pandemic-disruption-from-homework-gap-to-remote-learning-chasm/.

23. Javeria Salman, "How One City Closed the Digital Divide for Nearly All Its Students," *Hechinger Report*, April 14, 2022, https://hechingerreport.org/how-one-city-closed-the-digital-divide-for-nearly-all-its-students/.

24. Salman, "How One City Closed the Digital Divide."

25. Salman, "How One City Closed the Digital Divide."

26. Salman, "How One City Closed the Digital Divide."

27. Natalie Campisi, "Millions of Americans Are Still Missing Out on Broadband Access and Leaving Money on the Table—Here's Why," *Forbes Advisor*, May 26, 2023, https://www.forbes.com/advisor/personal-finance/millions-lack-broadband-access/.

28. US Government Accountability Office, "Closing the Digital Divide for the Millions of Americans Without Broadband," *WatchBlog*, February 1, 2023, https://www.gao.gov/blog/closing-digital-divide-millions-americans-without-broadband.

29. William Gibson, David Brin, and Anne Simon, "The Science in Science Fiction," interview by Brooke Gladstone, *Talk of the Nation*, NPR, November 30, 1999, radio broadcast, MP3 copy, 50:23, https://www.npr.org/2018/10/22/1067220/the-science-in-science-fiction.

30. Joy Buolamwini, "'If You Have a Face, You Have a Place in the Conversation About AI,' Expert Says," interview by Tonya Mosley, *Fresh Air*, NPR, November 28, 2023, audio, 37:58, https://www.npr.org/2023/11/28/1215529902/unmasking-ai-facial-recognition-technology-joy-buolamwini.

31. Buolamwini, "'If You Have a Face.'"

32. Joon Sung Park, Michael S. Bernstein, Robin N. Brewer, Ece Kamar, and Meredith Ringel Morris, "Understanding the Representation and Representativeness of Age in AI Data Sets," *AIES '21: Proceedings of the 2021 AAAI/ACM Conference on AI Ethics and Society* (2021): 834–842, https://doi.org/10.1145/3461702.3462590.

33. Joy Buolamwini and Timnit Gebru, "Gender Shades: Intersectional Accuracy Disparities in Commercial Gender Classification," *Proceedings of Machine*

Learning Research 81 (2018): 77–91, http://proceedings.mlr.press/v81/buolamwini18a/buolamwini18a.pdf.

34. Khari Johnson, "IBM Walked Away from Facial Recognition. What About Amazon and Microsoft?," *VentureBeat*, June 10, 2020, https://web.archive.org/web/20250416101000/https://venturebeat.com/ai/ibm-walked-away-from-facial-recognition-what-about-amazon-and-microsoft/ (site discontinued).

35. National Institute of Standards and Technology, "NIST Study Evaluates Effects of Race, Age, Sex on Face Recognition Software," press release, December 19, 2019, https://www.nist.gov/news-events/news/2019/12/nist-study-evaluates-effects-race-age-sex-face-recognition-software.

Prologue, Part II

1. Andrea Rupert et al., "It Takes a Village to Raise a Child: Understanding and Expanding the Concept of the 'Village,'" *Frontiers in Public Health* 10 (2022): 756066. https://doi.org/10.3389/fpubh.2022.756066.

2. "Automobile History," History.com, last updated February 27, 2025, https://www.history.com/articles/automobiles.

3. International Energy Agency, *Global EV Outlook 2024: Moving Towards Increased Affordability* (Paris: IEA Publications, 2024).

4. Arie J. Haagen-Smit, "Chemistry and Physiology of Los Angeles Smog," *Industrial and Engineering Chemistry* 44, no. 6 (1952): 1342–1346, https://web.archive.org/web/20170308044020/https://pdfs.semanticscholar.org/1f18/64424a52955b607781d45f007ed8fd1b8002.pdf.

5. Lev D. Labzovskii, Su-Jong Jeong, and Nicholas C. Parazoo, "Working Towards Confident Spaceborne Monitoring of Carbon Emissions from Cities Using Orbiting Carbon Observatory-2," *Remote Sensing of Environment* 233 (November 2019), https://doi.org/10.1016/j.rse.2019.111359.

6. Ralph Nader, *Unsafe at Any Speed: The Designed-In Dangers of the American Automobile* (New York: Grossman, 1965).

Chapter 3

1. Todd Litman, *Autonomous Vehicle Implementation Predictions: Implications for Transport Planning* (Victoria, BC: Victoria Transport Policy Institute, 2025), https://www.vtpi.org/avip.pdf.

2. Soumya Sudhakar, Vivienne Sze, and Sertac Karaman, "Data Centers on Wheels: Emissions from Computing Onboard Autonomous Vehicles," *IEEE Micro* 43, no. 1 (2023): 29–39, https://doi.org/10.1109/MM.2022.3219803.

3. Sudhakar, Sze, and Karaman, "Data Centers on Wheels."

4. Charlotte Elton, "Driverless Cars: The Dark Side of Autonomous Vehicles That No One's Talking About," *Euronews*, January 16, 2023, https://www.euronews.com/green/2023/01/16/driverless-cars-the-dark-side-of-autonomous-vehicles-that-no-ones-talking-about.

5. Sudhakar, Sze, and Karaman, "Data Centers on Wheels."

6. Wikipedia, "Agile Software Development," last modified August 27, 2025, 8:00 (UTC), https://en.wikipedia.org/wiki/Agile_software_development.

7. Kent Beck et al., "Manifesto for Agile Software Development," Agile Manifesto, 2001, https://agilemanifesto.org/.

8. Reggie Baker, "Agile vs Waterfall—Which Project Management Methodology to Choose?," *nTask* (blog), June 19, 2018, https://www.ntaskmanager.com/blog/agile-vs-waterfall/.

9. Jeff Sutherland, "How the Agile Manifesto Came to Be," posted February 11, 2021, by Scrum Inc., YouTube video, 15:06, https://www.youtube.com/watch?v=AGdi3zZutXk.

10. Project Management Institute, *Pulse of the Profession 2024: The Future of Project Work; Moving Past Office-Centric Models*, PMI, 2023, https://www.pmi.org/-/media/pmi/documents/public/pdf/learning/thought-leadership/pmi-pulse-of-the-profession-2024-report.pdf.

11. Emily Nutwell, email message to author, July 18, 2025.

12. Antonio Nieto-Rodriguez, "It's Time to End the Battle Between Waterfall and Agile," *Harvard Business Review*, October 10, 2023, https://hbr.org/2023/10/its-time-to-end-the-battle-between-waterfall-and-agile.

13. Jennifer Pahlka, *Recoding America: Why Government Is Failing in the Digital Age and How We Can Do Better* (New York: Metropolitan Books, 2023).

14. Vivek Wadhwa, "Code for America: An Elegant Solution for Government IT Problems," *Washington Post*, December 19, 2011, https://www.washingtonpost.com/national/on-innovations/code-for-america-an-elegant-solution-for-government-it

-problems/2011/12/16/gIQAXrIu2O_story.html; *CalFresh Program Overview*, presented to the Senate Human Services Committee and the Assembly Human Services Committee, March 11, 2014, California Legislative Analyst's Office, https://lao.ca.gov/handouts/socservices/2014/CalFresh-Overview-031114.pdf#:~:text=Rate%20Is%20Low.%20The%20United%20States%20Department,participation%20rate%20equal%20to%20that%20of%20California).

15. Pahlka, *Recoding America*, 136.

16. Pahlka, *Recoding America*, 174–175.

17. Code for America, "Simplifying California's Online Application for Food Benefits," accessed September 23, 2025, https://codeforamerica.org/success-stories/simplifying-californias-online-application-for-food-benefits/.

18. Grace Song and Zachary Antoyan, "Reflecting on 10 Years of Food Assistance in California," Code for America, August 27, 2024, https://codeforamerica.org/news/reflecting-on-10-years-of-getcalfresh/.

19. Kaitlyn Levinson, "Why California Is Revamping Its Benefits Application Websites," Route Fifty, December 20, 2024, https://www.route-fifty.com/customer-experience/2024/12/why-california-revamping-its-benefits-application-websites/401823/.

20. Code for America, "Simplifying California's Online Application for Food Benefits."

21. *Federal Motor Vehicle Safety Standards and Regulations*, National Highway Traffic Safety Administration (Washington, DC: US Department of Transportation, 1968), https://web.archive.org/web/20140529033515/http://www.nhtsa.gov/cars/rules/import/fmvss/.

22. David Drummond, "A New Approach to China," *Google Blog*, January 12, 2010, https://googleblog.blogspot.com/2010/01/new-approach-to-china.html.

23. Mat Honan, "How Apple and Amazon Security Flaws Led to My Epic Hacking," *Wired*, August 6, 2012, https://www.wired.com/2012/08/apple-amazon-mat-honan-hacking/.

24. Kevin Bankston, Ross Schulman, and Liz Woolery, "Case Study #2: Offering Two-Factor Authentication," in *Getting Internet Companies to Do the Right Thing* (Washington, DC: New America, 2017), https://www.newamerica.org/in-depth/getting-internet-companies-do-right-thing/case-study-2-offering-two-factor-authentication/.

25. Jen A. Miller, "Fern Hollow Collapse Illustrates Poor US Bridge Maintenance," *Construction Dive*, July 5, 2023, https://www.constructiondive.com/news/fern-hollow-collapse-illustrates-lack-of-bridge-maintenance-in-us/654194/.

26. American Society of Civil Engineers, *2025 Report Card for America's Infrastructure* (Reston, VA: American Society of Civil Engineers, 2025), https://infrastructurereportcard.org/wp-content/uploads/2025/03/Full-Report-2025-Natl-IRC-WEB.pdf; American Society of Civil Engineers, *2021 Report Card for America's Infrastructure* (Reston, VA: American Society of Civil Engineers, 2021), https://infrastructurereportcard.org/wp-content/uploads/2020/12/2021-IRC-Executive-Summary-1.pdf.

27. American Society of Civil Engineers, *2025 Report Card for America's Infrastructure*, 3.

28. Lee Vinsel and Andrew L. Russell, *The Innovation Delusion: How Our Obsession with the New Has Disrupted the Work That Matters Most* (New York: Crown Currency, 2020).

29. Vinsel and Russell, *Innovation Delusion*, front flap.

30. Vinsel and Russell, *Innovation Delusion*, 18.

31. Thoughtworks, *Responsible Technology Playbook: Tools for the United Nations*, United Nations and Thoughtworks, 2023, https://www.thoughtworks.com/en-us/insights/e-books/responsible-tech-playbook-un.

32. Thoughtworks, *Responsible Technology Playbook*.

33. UN System Chief Executives Board for Coordination, "Principles for the Ethical Use of Artificial Intelligence in the United Nations System," CEB/2022/2/Add.1, October 27, 2022, https://unsceb.org/sites/default/files/2023-03/CEB_2022_2_Add.1%20%28AI%20ethics%20principles%29.pdf.

Chapter 4

1. Aisha Playton, "Hip-Hop's Copyright Code: A Deep Dive into How the Genre Has Forged Its Own Ethic in IP," De Beer Attorneys, *IP, Decoded* (blog), updated October 27, 2024, https://www.debeerattorneys.com/post/hip-hop-copyright-ip.

2. Chuck D and Hank Shocklee, "How Copyright Law Changed Hip Hop: An Interview with Public Enemy's Chuck D and Hank Shocklee," interview by Kembrew McLeod, *Little Village*, October 17, 2011, https://littlevillagemag.com/how

-copyright-law-changed-hip-hop-an-interview-with-public-enemys-chuck-d-and-hank-shocklee/.

3. National Archives, "Medicare and Medicaid Act (1965)," Milestone Documents, last reviewed February 8, 2022, https://www.archives.gov/milestone-documents/medicare-and-medicaid-act.

4. US Environmental Protection Agency, *Clean Water Rule Comment Compendium: Mass Mailing Campaigns*, Response to Comments for the Clean Water Rule, June 2015, https://www.epa.gov/sites/default/files/2015-06/documents/cwr_response_to_comments_mass_mailing_campaigns.pdf.

5. *Last Week Tonight*, season 1, episode 5, "Net Neutrality," hosted by John Oliver, directed by John Perota, written by Ken Avery, Tim Carvell, and Dan Gurewich, aired June 1, 2014, on HBO, video, 13:17, June 2, 2014, https://www.youtube.com/watch?v=fpbOEoRrHyU.

6. Chris Jay Hoofnagle, Bart van der Sloot, and Frederik Zuiderveen Borgesius, "The European Union General Data Protection Regulation: What It Is and What It Means," *Information and Communications Technology Law* 28, no. 1 (2019): 65–98, https://doi.org/10.1080/13600834.2019.1573501.

7. Kristof Van Quathem, "European Commission Confirms Plans to Simplify GDPR," Covington, *Inside Privacy* (blog), March 17, 2025, https://www.insideprivacy.com/gdpr/european-commission-confirms-plans-to-simplify-gdpr/.

8. Hoofnagle, van der Sloot, and Zuiderveen Borgesius, "The European Union General Data Protection Regulation."

9. Ben Wolford, "What Is GDPR, the EU's New Data Protection Law?," GDPR.eu, accessed September 5, 2025, https://gdpr.eu/what-is-gdpr/.

10. European Union, "History of the European Union 1990–99," accessed September 5, 2025, https://european-union.europa.eu/principles-countries-history/history-eu/1990-99_en.

11. Hoofnagle, van der Sloot, and Zuiderveen Borgesius, "The European Union General Data Protection Regulation."

12. European Commission, "Digital Omnibus Regulation Proposal," *Shaping Europe's Digital Future*, Directorate-General for Communications Networks, Content and Technology, November 19, 2025, https://digital-strategy.ec.europa.eu/en/library/digital-omnibus-regulation-proposal.

13. Fredric Bellamy, "U.S. Data Privacy Laws to Enter New Era in 2023," *Reuters*, January 12, 2023, https://www.reuters.com/legal/legalindustry/us-data-privacy-laws-enter-new-era-2023-2023-01-12/.

14. Melissa Heikkilä, "Five Things You Need to Know About the EU's New AI Act," *MIT Technology Review*, December 11, 2023, https://www.technologyreview.com/2023/12/11/1084942/five-things-you-need-to-know-about-the-eus-new-ai-act/.

15. European Parliament, "EU AI Act: First Regulation on Artificial Intelligence," Topics, last updated February 19, 2025, https://www.europarl.europa.eu/topics/en/article/20230601STO93804/eu-ai-act-first-regulation-on-artificial-intelligence; Future of Life Institute, "High-Level Summary of the AI Act," EU Artificial Intelligence Act, last updated May 30, 2024, https://artificialintelligenceact.eu/high-level-summary/.

16. European Parliament, "EU AI Act."

17. Organisation for Economic Co-operation and Development, "Artificial Intelligence," accessed January 27, 2026, https://www.oecd.org/en/topics/policy-issues/artificial-intelligence.html.

18. White House, "Blueprint for an AI Bill of Rights," Office of Science and Technology Policy, accessed September 5, 2025, https://bidenwhitehouse.archives.gov/ostp/ai-bill-of-rights/.

19. White House, "Blueprint for an AI Bill of Rights."

20. Exec. Order No. 14110, 88 Fed. Reg. 75191 (October 30, 2023).

21. Exec. Order No. 14179, 90 Fed. Reg. 8741 (January 31, 2025).

22. Office of Science and Technology Policy, *Winning the Race: America's AI Action Plan*, Executive Office of the President of the United States, July 2025, https://www.whitehouse.gov/wp-content/uploads/2025/07/Americas-AI-Action-Plan.pdf.

23. Office of Science and Technology Policy, *Winning the Race*, 1.

24. Office of Science and Technology Policy, *Winning the Race*, 3.

25. Office of Science and Technology Policy, *Winning the Race*, 14–15.

26. James Kurose, "Who Controls the Internet?," online suppl. to chap. 1 of *Computer Networking*, by James F. Kurose and Keith W. Ross, 8th ed. (Hoboken, NJ: Pearson, 2020), posted February 14, 2022, YouTube video, 21:51, https://www.youtube.com/watch?v=xrd4hD_9fS8.

27. Gail Sullivan, "What the Heck Is Net Neutrality?," *Washington Post*, May 15, 2014, https://www.washingtonpost.com/news/morning-mix/wp/2014/05/15/everything-you-need-to-know-about-net-neutrality/.

28. James F. Kurose and Keith W. Ross, *Computer Networking: A Top-Down Approach*, 8th ed. (New York: Pearson, 2020).

29. Federal Communications Commission, "Protecting and Promoting the Open Internet," *Federal Register* 80, no. 70 (April 13, 2015): 19,738–19,801.

30. Tim Wu, *The Master Switch: The Rise and Fall of Information Empires* (New York: Knopf, 2010).

31. Tim Walker, "Take Cellphones Out of the Classroom, Educators Say," *NEA Today*, October 3, 2024, https://www.nea.org/nea-today/all-news-articles/take-cellphones-out-classroom-educators-say.

32. Arianna Prothero, Lauraine Langreo, and Alyson Klein, "Which States Ban or Restrict Cellphones in Schools?," *Education Week*, updated July 17, 2025, https://www.edweek.org/technology/which-states-ban-or-restrict-cellphones-in-schools/2024/06.

33. NEA Task Force on Artificial Intelligence in Education, *Report of the NEA Task Force on Artificial Intelligence in Education* (Washington, DC: National Education Association, 2024), https://www.nea.org/sites/default/files/2024-10/nea-ai-task-force-report-2024.pdf.

34. Lydia Gichuki, "AI Goes to School: The Global AI Education Race, Opportunities and Perils," DevelopmentAid, April 30, 2025, https://www.developmentaid.org/news-stream/post/194647/ai-transforming-education.

Chapter 5

1. *Facebook, Social Media Privacy, and the Use and Abuse of Data: Joint Hearing Before the S. Comm. on Com., Sci., and Transp. and the S. Comm. on the Judiciary*, 115th Cong. 21 (2018) (statement of Senator Hatch).

2. *Facebook, Social Media Privacy, and the Use and Abuse of Data: Joint Hearing Before the S. Comm. on Com., Sci., and Transp. and the S. Comm. on the Judiciary*, 115th Cong. 96 (2018) (statement of Senator Schatz).

3. Emily Stewart, "Lawmakers Seem Confused About What Facebook Does—and How to Fix It," *Vox*, April 10, 2018, https://www.vox.com/policy-and-politics/2018/4/10/17222062/mark-zuckerberg-testimony-graham-facebook-regulations.

4. "Tech for the American People," US Tech Force, accessed January 2, 2026, https://techforce.gov.

5. Natalie Alms, "IRS Is Making Headway on Modernizing 1960s-Era Tax System, Commissioner Says," *NextGov/FCW*, May 20, 2024, https://www.nextgov.com/modernization/2024/05/irs-making-headway-modernizing-1960s-era-tax-system-commissioner-says/396695/.

6. David Jensen, "PIT and the OTA," lecture, University of Massachusetts Amherst, Amherst, MA, April 14, 2025.

7. *Budget Request to the Office of Management and Budget: Fiscal Year 1995* (Washington, DC: Peace Corps, 1995), https://files.peacecorps.gov/manuals/cbj/1995.pdf?_gl=1*9yola2*_gcl_au*NTE3NjQ5NzU5LjE3NzI4MDQxMDI.*_ga*MTI1MzE4MjU3My4xNzcyODA0MTAy*_ga_7WoV5D49XM*czE3NzI4MDQxMDIkbzEkZzEkdDE3NzI4MDQ0ODkkajM1JGwwJGgw.

8. Harry McCracken, "1995: The Year Everything Changed," *Fast Company*, December 30, 2015, https://www.fastcompany.com/3053055/1995-the-year-everything-changed.

9. Gregory C. Kunkle, "New Challenge or the Past Revisited? The Office of Technology Assessment in Historical Context," *Technology in Society* 17, no. 2 (1995): 175–196, https://doi.org/10.1016/0160-791X(95)00007-E.

10. Jensen, "PIT and the OTA"; Emily G. Blevins, *The Office of Technology Assessment: History, Authorities, Issues, and Options*, Congressional Research Service, CRS Report R46327 (Washington, DC: Office of Congressional Information and Publishing, April 29, 2020), https://www.congress.gov/crs_external_products/R/PDF/R46327/R46327.8.pdf.

11. Blevins, *Office of Technology Assessment*.

12. *Who Is Selling Your Data: A Critical Examination of the Role of Data Brokers in the Digital Economy; Hearing Before the Subcomm. on Oversight and Investigations of the H. Comm. on Energy and Commerce*, 118 Cong. 6 (2023) (testimony of Justin Sherman, Senior Fellow and Research Lead, Data Brokerage Project, Duke University Sanford School of Public Policy).

13. Electronic Privacy Information Center, "Data Brokers," accessed September 7, 2025, https://epic.org/issues/consumer-privacy/data-brokers/.

14. Douglas MacMillan, "Data Brokers Are Selling Your Secrets. How States Are Trying to Stop Them," *Washington Post*, June 24, 2019, https://www.washingtonpost.com/business/2019/06/24/data-brokers-are-getting-rich-by-selling-your-secrets-how-states-are-trying-stop-them/.

15. Matt Burgess, "Amazon's Massive GDPR Fine Shows the Law's Power—and Limits," *Wired*, August 5, 2021, https://www.wired.com/story/amazon-gdpr-fine-shows-power-limits/.

16. Adam Satariano and Jenny Gross, "Google and Apple Face Billions in Penalties After Losing E.U. Appeals," *New York Times*, September 10, 2024, https://www.nytimes.com/2024/09/10/technology/european-union-apple-google-antitrust.html.

17. "20 Biggest GDPR Fines So Far [2025]," *Data Privacy Manager* (blog), March 3, 2025, https://dataprivacymanager.net/5-biggest-gdpr-fines-so-far-2020/.

18. Chris Jay Hoofnagle, Bart van der Sloot, and Frederik Zuiderveen Borgesius, "The European Union General Data Protection Regulation: What It Is and What It Means," *Information and Communications Technology Law* 28, no. 1 (2019): 65–98, https://doi.org/10.1080/13600834.2019.1573501.

19. Paul Mozur, Raymond Zhong, and Aaron Krolik, "In Coronavirus Fight, China Gives Citizens a Color Code, With Red Flags," *New York Times*, March 1, 2020, https://www.nytimes.com/2020/03/01/business/china-coronavirus-surveillance.html; Nectar Gan and David Culver, "China Is Fighting the Coronavirus with a Digital QR Code. Here's How It Works," *CNN*, April 16, 2020, https://www.cnn.com/2020/04/15/asia/china-coronavirus-qr-code-intl-hnk/index.html.

20. WHO–China Joint Mission on Coronavirus Disease 2019, *Report of the WHO–China Joint Mission on Coronavirus Disease 2019 (COVID-19)*, World Health Organization, February 28, 2020, https://www.who.int/docs/default-source/coronaviruse/who-china-joint-mission-on-Covid-19-final-report.pdf.

21. Ken (Jianmin) Dai and Jet (Zhisong) Deng, "China's Personal Information Protection Law (PIPL)," Bloomberg Law, April 12, 2022, https://pro.bloomberglaw.com/insights/privacy/china-personal-information-protection-law-pipl-faqs/.

22. Maureen Fan, "China's Party Leadership Declares New Priority: 'Harmonious Society,'" *Washington Post*, October 12, 2006, https://www.washingtonpost.com/wp-dyn/content/article/2006/10/11/AR2006101101610.html.

Chapter 6

1. European Union Agency for Fundamental Rights, "EU Charter of Fundamental Rights: Article 8—Protection of Personal Data," *Official Journal of the European Union* C 303/17 (December 14, 2007), https://fra.europa.eu/en/eu-charter/article/8-protection-personal-data.

2. Kashmir Hill, "Max Schrems: The Austrian Thorn in Facebook's Side," *Forbes*, February 7, 2012, https://www.forbes.com/sites/kashmirhill/2012/02/07/the-austrian-thorn-in-facebooks-side/.

3. Max Schrems, Europe Versus Facebook (website), accessed September 7, 2025, http://www.europe-v-facebook.org/.

4. Derek Scally, "Max Schrems Files First Cases Under GDPR Against Facebook and Google," *Irish Times*, May 25, 2018, https://www.irishtimes.com/business/technology/max-schrems-files-first-cases-under-gdpr-against-facebook-and-google-1.3508177.

5. Emily Benson and Elizabeth Duncan, "Temporarily Shielded? Executive Action and the Transatlantic Data Privacy Framework," Center for Strategic and International Studies (CSIS), October 7, 2022, https://www.csis.org/analysis/temporarily-shielded-executive-action-and-transatlantic-data-privacy-framework.

6. NOYB, "Fines Resulting from NOYB Litigation," accessed September 7, 2025, https://noyb.eu/en/fines-resulting-noyb-litigation.

7. NOYB (website), accessed September 7, 2025, https://noyb.eu/en.

8. NOYB, "Fines Resulting from NOYB Litigation."

9. Julia Fioretti, Shadia Nasralla, and Francois Murphy, "Max Schrems: The Law Student Who Took On Facebook," *Reuters*, October 7, 2015, https://www.reuters.com/article/technology/max-schrems-the-law-student-who-took-on-facebook-idUSKCN0S1240/.

10. Rachel Carson, *Silent Spring* (Boston: Houghton Mifflin, 1962).

11. Chip Heath and Dan Heath, *Switch: How to Change Things When Change Is Hard* (New York: Broadway Books, 2010).

12. Heath and Heath, *Switch*, 255.

13. Federal Communications Commission, "Protecting and Promoting the Open Internet," *Federal Register* 79, no. 126 (July 1, 2014): 37448–37483.

14. *Last Week Tonight*, season 1, episode 5, "Net Neutrality," hosted by John Oliver, directed by John Perota, written by Ken Avery, Tim Carvell, and Dan Gurewich, aired June 1, 2014, on HBO, video, 13:17, June 2, 2014, https://www.youtube.com/watch?v=fpbOEoRrHyU; Meredith Blake, "John Oliver Rails Against Cable Companies over Net Neutrality," *LA Times*, June 2, 2014, https://www.latimes.com/entertainment/tv/showtracker/la-et-st-john-oliver-rails-against-cable-companies-over-net-neutrality-20140602-story.html.

15. Doug Aamoth, "John Oliver's Net Neutrality Rant Crashes FCC Servers," *Time*, June 3, 2014, https://time.com/2817567/john-oliver-net-neutrality-fcc/.

16. *Last Week Tonight*, season 4, episode 11, "Net Neutrality II," hosted by John Oliver, directed by Paul Pennolino, written by Jeff Maurer et al., aired May 7, 2017, on HBO, video, 19:32, May 8, 2017, https://www.youtube.com/watch?v=92vuuZt7wak.

17. Jon Brodkin, "FCC's Ajit Pai Says Net Neutrality's 'Days Are Numbered' Under Trump," *Ars Technica*, December 8, 2016, https://arstechnica.com/information-technology/2016/12/fccs-ajit-pai-says-net-neutralitys-days-are-numbered-under-trump/.

18. Gregory Wallace, "That Was No Cyberattack on the FCC, Inspector General Says—Just John Oliver Fans," *CNN*, August 8, 2018, https://www.cnn.com/2018/08/08/politics/fcc-website-crash-ig-john-oliver-show-hbo-net-neutrality/index.html.

19. Wallace, "That Was No Cyberattack."

20. Ashley McBride, "Oakland Made Gains in Closing the Digital Divide During COVID. The Classroom Tech Could Be Here to Stay," *Oaklandside*, July 20, 2021, https://oaklandside.org/2021/07/20/oakland-made-gains-in-closing-the-digital-divide-during-covid-the-classroom-tech-could-be-here-to-stay/.

21. City of Oakland, "City of Oakland Rolls Out Broadband Master Plan," June 12, 2025, https://www.oaklandca.gov/News-Releases/City-of-Oakland-Rolls-Out-Broadband-Master-Plan.

22. Heath and Heath, *Switch*.

23. Deloitte, *2024 Gen Z and Millennial Survey*, Deloitte, May 2024, https://www.eticanews.it/wp-content/uploads/2024/05/deloitte-2024-genz-millennial-survey.pdf, 3.

24. Public Interest Technology University Network, "Our Work," accessed January 23, 2026, https://pit-un.org/about-us/.

25. Wikipedia, "Public Interest Technology University Network," last modified August 26, 2025, 21:14 (UTC), https://en.wikipedia.org/wiki/Public_Interest_Technology_University_Network.

26. UMass Amherst, "Public Interest Technology Initiative," accessed September 7, 2025, https://groups.cs.umass.edu/pit/.

Chapter 7

1. Shoshana Zuboff, *The Age of Surveillance Capitalism: The Fight for a Human Future at the New Frontier of Power* (London: Profile Books, 2019).

2. Alexandra Borgeaud, "Surveillance Technology Market Size Worldwide from 2022 to 2027," infographic, Statista, February 28, 2025, https://www.statista.com/statistics/1251839/surveillance-technology-market-global/.

3. Douglas MacMillan, David Ovalle, and Aaron Schaffer, "Arrested by AI: Police Ignore Standards After Facial Recognition Matches," *Washington Post*, January 13, 2025, https://www.washingtonpost.com/business/interactive/2025/police-artificial-intelligence-facial-recognition/.

4. Gallup, "Global Indicator: Hybrid Work," Gallup, Inc., accessed January 2, 2026, https://www.gallup.com/401384/indicator-hybrid-work.aspx.

5. Jasper Hamill, "Big Boss Is Watching You: The Growth of Workplace Surveillance Technology," *Stack*, September 12, 2024, https://www.thestack.technology/workplace-digital-surveillance/.

6. US Government Accountability Office, "'Why Do I Feel like Somebody's Watching Me?' Workplace Surveillance Can Impact More than Just Productivity," *WatchBlog*, October 29, 2024, https://www.gao.gov/blog/why-do-i-feel-somebodys-watching-me-workplace-surveillance-can-impact-more-just-productivity.

7. World Economic Forum, *Future of Jobs Report 2025* (Geneva: World Economic Forum, January 2025), https://reports.weforum.org/docs/WEF_Future_of_Jobs_Report_2025.pdf, 26.

8. Alexis de Tocqueville, *Journeys to England and Ireland (1835)* (New Haven, CT: Yale University Press, 1958), 104–107, https://sites.pitt.edu/~syd/toq.html#:~:text=104%2D7.,and%20over%20the%20whole%20world.

9. World Economic Forum, *Future of Jobs Report.*

10. World Economic Forum, *Future of Jobs Report.*

11. World Economic Forum, *Future of Jobs Report.*

12. Bernard Marr, "How Generative AI Will Change the Jobs of Architects and Civil Engineers," *Forbes*, March 15, 2024, https://www.forbes.com/sites/bernard marr/2024/03/15/how-generative-ai-will-change-the-jobs-of-architects-and-civil -engineers/.

13. Statista, "Countries with Most Content Available on Netflix Worldwide as of February 2025," infographic, Statista, February 2025, https://www.statista.com /statistics/1013571/netflix-library-size-worldwide/.

14. Stuart Russell and Peter Norvig, *Artificial Intelligence: A Modern Approach*, 4th ed., global ed. (Hoboken, NJ: Pearson, 2021).

15. Stuart Russell, "Beyond ChatGPT: Stuart Russell on the Risks and Rewards of A.I.," interview by Jerry Kaplan, Commonwealth Club of California, San Francisco, CA, April 3, 2023, YouTube video, 1:13:24, https://www.youtube.com/watch?v=ow3 XrwTmFA8.

Chapter 8

1. *2001: A Space Odyssey*, directed by Stanley Kubrick (Metro-Goldwyn-Mayer, 1968).

2. Stuart Russell and Peter Norvig, *Artificial Intelligence: A Modern Approach*, 4th ed., global ed. (Hoboken, NJ: Pearson, 2021).

3. Russell and Norvig, *Artificial Intelligence*, chap. 1.

4. Katja Grace et al., "Thousands of AI Authors on the Future of AI," *Journal of Artificial Intelligence Research* 84 (2025), https://doi.org/10.1613/jair.1.19087.

5. Russell and Norvig, *Artificial Intelligence*, 51.

6. Josh Simons, *Algorithms for the People: Democracy in the Age of AI* (Princeton, NJ: Princeton University Press, 2023), 14.

7. Cathy O'Neil, *Weapons of Math Destruction: How Big Data Increases Inequality and Threatens Democracy* (New York: Crown, 2016), 21.

8. Emily M. Bender et al., "On the Dangers of Stochastic Parrots: Can Language Models Be Too Big?," *FAccT '21: Proceedings of the 2021 ACM Conference on Fairness,*

Accountability, and Transparency (2021): 616–617, https://doi.org/10.1145/3442188.3445922.

9. Cade Metz, "Silicon Valley's Elusive Fantasy of a Computer as Smart as You," *New York Times*, May 16, 2025, https://www.nytimes.com/2025/05/16/technology/what-is-agi.html.

10. Matteo Wong, "The GPT Era Is Already Ending," *The Atlantic*, December 6, 2024, https://www.theatlantic.com/technology/archive/2024/12/openai-o1-reasoning-models/680906/.

11. Stuart Russell, "Beyond ChatGPT: Stuart Russell on the Risks and Rewards of A.I.," interview by Jerry Kaplan, Commonwealth Club of California, San Francisco, CA, April 3, 2023, video, 1:13:24, https://www.youtube.com/watch?v=ow3XrwTmFA8.

12. Wikipedia, "Three Laws of Robotics," last modified February 3, 2026, 2:10 (UTC), https://en.wikipedia.org/wiki/Three_Laws_of_Robotics.

13. Frank Pasquale, *New Laws of Robotics: Defending Human Expertise in the Age of AI* (Cambridge, MA: Belknap Press, 2020), 3–12.

14. Kate Darling, *The New Breed: What Our History with Animals Reveals about Our Future with Robots* (New York: Henry Holt, 2021).

15. Darling, *New Breed*.

16. Darling, *New Breed*, 82.

17. Darling, *New Breed*, 74.

18. Darling, *New Breed*, 60–61.

19. Simons, *Algorithms for the People*, 7.

20. Simons, *Algorithms for the People*, 181.

21. Simons, *Algorithms for the People*, 181.

22. Simons, *Algorithms for the People*, 186–187.

23. Tim Wu, *The Master Switch: The Rise and Fall of Information Empires* (New York: Knopf, 2010).

24. Simons, *Algorithms for the People*, 190.

25. American Civil Liberties Union, "ACLU v. Clearview AI," last updated May 11, 2022, https://www.aclu.org/cases/aclu-v-clearview-ai.

26. Exec. Order No. 14110, 88 Fed. Reg. 75191 (October 30, 2023); Exec. Order No. 14179, 90 Fed. Reg. 8741 (January 31, 2025).

27. Nathalie A. Smuha, "Concluding Remarks," in *The Cambridge Handbook of the Law, Ethics and Policy of Artificial Intelligence*, ed. Nathalie A. Smuha (Cambridge: Cambridge University Press, 2025), 430.

28. Kiel Brennan-Marquez, personal communication, February 11, 2025.

29. Abeba Birhane, Jelle van Dijk, and Frank Pasquale, "Debunking Robot Rights Metaphysically, Ethically, and Legally," *First Monday* 29, no. 4 (April 2024): abstract, https://doi.org/10.5210/fm.v29i4.13628.

30. Smuha, "Concluding Remarks," 430.

Chapter 9

1. Alexander Nix, "The Power of Big Data and Psychographics," presentation, 2016 Concordia Annual Summit, New York, NY, September 19, 2016, posted September 27, 2016, YouTube video, 11:00, https://www.youtube.com/watch?v=n8Dd5aVXLCc.

2. Nix, "The Power of Big Data."

3. Donald A. Norman, *The Psychology of Everyday Things* (New York: Basic Books, 1988).

4. Taylor Majewski, "It's Time to Retire the Term 'User,'" *MIT Technology Review*, April 19, 2024, https://www.technologyreview.com/2024/04/19/1090872/ai-users-people-terms/.

5. Donald A. Norman, "Words Matter. Talk About People: Not Customers, Not Consumers, Not Users," JND.org, November 17, 2008, https://jnd.org/words-matter-talk-about-people-not-customers-not-consumers-not-users/.

6. Siddhartha Mukherjee, "Cancer's Invasion Equation," *New Yorker*, September 4, 2017, https://www.newyorker.com/magazine/2017/09/11/cancers-invasion-equation.

7. Mukherjee, "Cancer's Invasion Equation."

8. Knight First Amendment Institute, "Federal Judge Says Challenge in Zuckerman v. Meta Cannot Go Forward Yet," press release, November 8, 2024, https://knightcolumbia.org/content/federal-judge-says-challenge-in-zuckerman-v-meta-cannot-go-forward-yet.

9. Mukherjee, "Cancer's Invasion Equation."

Appendix

1. Emily M. Bender et al., "On the Dangers of Stochastic Parrots: Can Language Models Be Too Big?," *FAccT '21: Proceedings of the 2021 ACM Conference on Fairness, Accountability, and Transparency* (2021): 610–623, https://doi.org/10.1145/3442188.3445922.

BIBLIOGRAPHY

Aamoth, Doug. "John Oliver's Net Neutrality Rant Crashes FCC Servers." *Time*, June 3, 2014. https://time.com/2817567/john-oliver-net-neutrality-fcc/.

Alms, Natalie. "IRS Is Making Headway on Modernizing 1960s-Era Tax System, Commissioner Says." *NextGov/FCW*, May 20, 2024. https://www.nextgov.com/modernization/2024/05/irs-making-headway-modernizing-1960s-era-tax-system-commissioner-says/396695/.

American Civil Liberties Union. "ACLU v. Clearview AI." Updated May 11, 2022. https://www.aclu.org/cases/aclu-v-clearview-ai.

American Society of Civil Engineers. *2021 Report Card for America's Infrastructure*. Reston, VA: American Society of Civil Engineers, 2021. https://infrastructurereportcard.org/wp-content/uploads/2020/12/2021-IRC-Executive-Summary-1.pdf.

American Society of Civil Engineers. *2025 Report Card for America's Infrastructure*. Reston, VA: American Society of Civil Engineers, 2025. https://infrastructurereportcard.org/wp-content/uploads/2025/03/Full-Report-2025-Natl-IRC-WEB.pdf.

"Automobile History." History.com, last updated February 27, 2025. https://www.history.com/articles/automobiles.

Baker, Reggie. "Agile vs Waterfall—Which Project Management Methodology to Choose?" *nTask* (blog), June 19, 2018. https://www.ntaskmanager.com/blog/agile-vs-waterfall/.

Bankston, Kevin, Ross Schulman, and Liz Woolery. "Case Study #2: Offering Two-Factor Authentication." In *Getting Internet Companies to Do the Right Thing*. Washington, DC: New America, 2017. https://www.newamerica.org/in-depth/getting-internet-companies-do-right-thing/case-study-2-offering-two-factor-authentication/.

Beck, Kent, Mike Beedle, Arie van Bennekum et al. "Manifesto for Agile Software Development." Agile Manifesto, 2001. https://agilemanifesto.org/.

Bellamy, Fredric. "U.S. Data Privacy Laws to Enter New Era in 2023." *Reuters*, January 12, 2023. https://www.reuters.com/legal/legalindustry/us-data-privacy-laws-enter-new-era-2023-2023-01-12/.

Bender, Emily M., Timnit Gebru, Angelina McMillan-Major, and Shmargaret Shmitchell. "On the Dangers of Stochastic Parrots: Can Language Models Be Too Big?" *FAccT '21: Proceedings of the 2021 ACM Conference on Fairness, Accountability, and Transparency* (2021): 610–623. https://doi.org/10.1145/3442188.3445922.

Benson, Emily, and Elizabeth Duncan. "Temporarily Shielded? Executive Action and the Transatlantic Data Privacy Framework." Center for Strategic and International Studies (CSIS), October 7, 2022. https://www.csis.org/analysis/temporarily-shielded-executive-action-and-transatlantic-data-privacy-framework.

BioSpace. "FDA Recalls 465,000 St. Jude Medical Pacemakers over Hacking Fears." Press release, August 31, 2017. https://www.biospace.com/fda-recalls-465-000-st-jude-medical-pacemakers-over-hacking-fears.

Birhane, Abeba, Jelle van Dijk, and Frank Pasquale. "Debunking Robot Rights Metaphysically, Ethically, and Legally." *First Monday* 29, no. 4 (April 2024). https://doi.org/10.5210/fm.v29i4.13628.

Blake, Meredith. "John Oliver Rails Against Cable Companies over Net Neutrality." *LA Times*, June 2, 2014. https://www.latimes.com/entertainment/tv/showtracker/la-et-st-john-oliver-rails-against-cable-companies-over-net-neutrality-20140602-story.html.

Blevins, Emily G. *The Office of Technology Assessment: History, Authorities, Issues, and Options*. Congressional Research Service. CRS Report R46327. Washington, DC: Office of Congressional Information and Publishing, April 29, 2020. https://www.congress.gov/crs_external_products/R/PDF/R46327/R46327.8.pdf.

Bohannon, Molly. "Lawyer Used ChatGPT in Court—and Cited Fake Cases. A Judge Is Considering Sanctions." *Forbes*, June 8, 2023. https://www.forbes.com/sites

/mollybohannon/2023/06/08/lawyer-used-chatgpt-in-court-and-cited-fake-cases-a-judge-is-considering-sanctions/.

Bond, Shannon. "A Must for Millions, Zoom Has a Dark Side—and an FBI Warning." *NPR*, April 3, 2020. https://www.npr.org/2020/04/03/826129520/a-must-for-millions-zoom-has-a-dark-side-and-an-fbi-warning.

Borgeaud, Alexandra. "Surveillance Technology Market Size Worldwide from 2022 to 2027." Infographic. Statista, February 28, 2025. https://www.statista.com/statistics/1251839/surveillance-technology-market-global/.

Brodkin, Jon. "FCC's Ajit Pai Says Net Neutrality's 'Days Are Numbered' Under Trump." *Ars Technica*, December 8, 2016. https://arstechnica.com/information-technology/2016/12/fccs-ajit-pai-says-net-neutralitys-days-are-numbered-under-trump/.

Buolamwini, Joy. "'If You Have a Face, You Have a Place in the Conversation About AI,' Expert Says." Interview by Tonya Mosley. *Fresh Air*, NPR, November 28, 2023. Audio, 37:58. https://www.npr.org/2023/11/28/1215529902/unmasking-ai-facial-recognition-technology-joy-buolamwini.

Buolamwini, Joy, and Timnit Gebru. "Gender Shades: Intersectional Accuracy Disparities in Commercial Gender Classification." *Proceedings of Machine Learning Research* 81 (2018): 77–91. http://proceedings.mlr.press/v81/buolamwini18a/buolamwini18a.pdf.

Burgess, Matt. "Amazon's Massive GDPR Fine Shows the Law's Power—and Limits." *Wired*, August 5, 2021. https://www.wired.com/story/amazon-gdpr-fine-shows-power-limits/.

Calabrese, Michael, and Amir Nasr. "Pandemic Disruption: From Homework Gap to Remote Learning Chasm." In *The Online Learning Equity Gap*. Washington, DC: New America, 2020. https://www.newamerica.org/oti/reports/online-learning-equity-gap/i-pandemic-disruption-from-homework-gap-to-remote-learning-chasm/.

Campisi, Natalie. "Millions of Americans Are Still Missing Out on Broadband Access and Leaving Money on the Table—Here's Why." *Forbes Advisor*, May 26, 2023. https://www.forbes.com/advisor/personal-finance/millions-lack-broadband-access/.

Carson, Rachel. *Silent Spring*. Boston: Houghton Mifflin, 1962.

Carville, Olivia. "Scammers Are Targeting Teenage Boys on Social Media—and Driving Some to Suicide." *Bloomberg*, April 15, 2024. https://www.bloomberg.com/features/2024-sextortion-teen-suicides/.

City of Oakland. "City of Oakland Rolls Out Broadband Master Plan." June 12, 2025. https://www.oaklandca.gov/News-Releases/City-of-Oakland-Rolls-Out-Broadband-Master-Plan.

Chuck D and Hank Shocklee. "How Copyright Law Changed Hip Hop: An Interview with Public Enemy's Chuck D and Hank Shocklee." Interview by Kembrew McLeod. *Little Village*, October 17, 2011. https://littlevillagemag.com/how-copyright-law-changed-hip-hop-an-interview-with-public-enemys-chuck-d-and-hank-shocklee/.

Code for America. "Simplifying California's Online Application for Food Benefits." Accessed September 23, 2025. https://codeforamerica.org/success-stories/simplifying-californias-online-application-for-food-benefits/.

Cybersecurity and Infrastructure Security Agency. "Critical Infrastructure Sectors." Accessed September 4, 2025. https://www.cisa.gov/topics/critical-infrastructure-security-and-resilience/critical-infrastructure-sectors.

Cybersecurity and Infrastructure Security Agency. "Cyber Threats to Medical Technology and Communication Technology Protocols." Infographic. CISA, February 2024. https://www.cisa.gov/sites/default/files/2024-02/Cyber%20Threats%20to%20Medical%20Technology%20and%20Communication%20Technology%20Protocols%20%28508%29.pdf.

Dahl, Ronald E., Nicholas B. Allen, Linda Wilbrecht, and Ahna Ballonoff Suleiman. "Importance of Investing in Adolescence from a Developmental Science Perspective." *Nature* 554 (2018): 441–450. https://doi.org/10.1038/nature25770.

Dai, Ken (Jianmin), and Jet (Zhisong) Deng. "China's Personal Information Protection Law (PIPL)." Bloomberg Law, April 12, 2022. https://pro.bloomberglaw.com/insights/privacy/china-personal-information-protection-law-pipl-faqs/.

Darling, Kate. *The New Breed: What Our History with Animals Reveals about Our Future with Robots*. New York: Henry Holt, 2021.

"Data: Coronavirus and School Closures in 2019–2020." *Education Week*, updated January 10, 2022. https://www.edweek.org/data-coronavirus-and-school-closures-in-2019-2020/2021/12.

Data Privacy Manager. "20 Biggest GDPR Fines So Far [2025]." *Data Privacy Manager* (blog), March 3, 2025. https://dataprivacymanager.net/5-biggest-gdpr-fines-so-far-2020/.

Deloitte. *2024 Gen Z and Millennial Survey.* Deloitte, May 2024. https://www.eticanews.it/wp-content/uploads/2024/05/deloitte-2024-genz-millennial-survey.pdf.

Drummond, David. "A New Approach to China." *Google Blog,* January 12, 2010. https://googleblog.blogspot.com/2010/01/new-approach-to-china.html.

Dunleavy, Brian P. "Pacemakers, Insulin Pumps Can Be Hacked, Experts Say." *United Press International,* June 1, 2022. https://www.upi.com/Health_News/2022/06/01/medical-devices-pacemakers-cybersecurity/7041653656330/.

Electronic Privacy Information Center. "Data Brokers." Accessed September 7, 2025. https://epic.org/issues/consumer-privacy/data-brokers/.

Elton, Charlotte. "Driverless Cars: The Dark Side of Autonomous Vehicles That No One's Talking About." *Euronews,* January 16, 2023. https://www.euronews.com/green/2023/01/16/driverless-cars-the-dark-side-of-autonomous-vehicles-that-no-ones-talking-about.

Estes, Adam Clark. "Why the Internet (Probably) Won't Break During the Coronavirus Pandemic." *Vox,* March 25, 2020. https://www.vox.com/recode/2020/3/25/21188391/internet-surge-traffic-coronavirus-pandemic.

European Commission. "Digital Omnibus Regulation Proposal." *Shaping Europe's Digital Future,* Directorate-General for Communications Networks, Content and Technology, November 19, 2025. https://digital-strategy.ec.europa.eu/en/library/digital-omnibus-regulation-proposal.

European Parliament. "EU AI Act: First Regulation on Artificial Intelligence." Topics. Last updated February 19, 2025. https://www.europarl.europa.eu/topics/en/article/20230601STO93804/eu-ai-act-first-regulation-on-artificial-intelligence.

European Union. "History of the European Union 1990–99." Accessed September 5, 2025. https://european-union.europa.eu/principles-countries-history/history-eu/1990-99_en.

European Union Agency for Fundamental Rights. "EU Charter of Fundamental Rights: Article 8—Protection of Personal Data." *Official Journal of the*

European Union C 303/17 (December 14, 2007). https://fra.europa.eu/en/eu-charter/article/8-protection-personal-data.

Executive Order No. 14110. 88 Federal Register 75191. "Safe, Secure, and Trustworthy Development and Use of Artificial Intelligence." October 30, 2023. https://www.federalregister.gov/documents/2023/11/01/2023-24283/safe-secure-and-trustworthy-development-and-use-of-artificial-intelligence.

Executive Order No. 14179. 90 Federal Register 8741. "Removing Barriers to American Leadership in Artificial Intelligence." January 31, 2025. https://www.federalregister.gov/documents/2025/01/31/2025-02172/removing-barriers-to-american-leadership-in-artificial-intelligence.

Fan, Maureen. "China's Party Leadership Declares New Priority: 'Harmonious Society.'" *Washington Post*, October 12, 2006. https://www.washingtonpost.com/wp-dyn/content/article/2006/10/11/AR2006101101610.html.

Federal Bureau of Investigation. "Sextortion: A Growing Threat Preying Upon Our Nation's Teens." Press release, January 17, 2024. https://www.fbi.gov/contact-us/field-offices/sacramento/news/sextortion-a-growing-threat-preying-upon-our-nations-teens.

Federal Communications Commission. "Protecting and Promoting the Open Internet." *Federal Register* 79, no. 126 (July 1, 2014): 37448–37483.

Federal Communications Commission. "Protecting and Promoting the Open Internet." *Federal Register* 80, no. 70 (April 13, 2015): 19,738–19,801.

Federal Motor Vehicle Safety Standards and Regulations. National Highway Traffic Safety Administration. Washington, DC: US Department of Transportation, 1968. https://web.archive.org/web/20140529033515/http://www.nhtsa.gov/cars/rules/import/fmvss/.

Federal Trade Commission. "Behind the FTC's Inquiry into Surveillance Pricing Practices." *FTC Office of Technology Blog*, July 23, 2024. https://www.ftc.gov/policy/advocacy-research/tech-at-ftc/2024/07/behind-ftcs-inquiry-surveillance-pricing-practices.

Fioretti, Julia, Shadia Nasralla, and Francois Murphy. "Max Schrems: The Law Student Who Took On Facebook." *Reuters*, October 7, 2015. https://www.reuters.com/article/technology/max-schrems-the-law-student-who-took-on-facebook-idUSKCN0S1240/.

Fornell, Dave. "FDA Confirms Cybersecurity Vulnerabilities of St. Jude's Implantable Cardiac Devices, Merlin Transmitter." *Diagnostic and Interventional Cardiology*, January 9, 2017. https://www.dicardiology.com/article/fda-confirms-cybersecurity-vulnerabilities-st-judes-implantable-cardiac-devices-merlin.

Fung, Brian. "We Finally Know What Caused the Global Tech Outage—and How Much It Cost." *CNN*, July 24, 2024. https://www.cnn.com/2024/07/24/tech/crowdstrike-outage-cost-cause/index.html.

Future of Life Institute. "High-Level Summary of the AI Act." EU Artificial Intelligence Act, last updated May 30, 2024. https://artificialintelligenceact.eu/high-level-summary/.

Future of Life Institute. "Pause Giant AI Experiments: An Open Letter." March 22, 2023. https://futureoflife.org/open-letter/pause-giant-ai-experiments/.

Gallup. "Global Indicator: Hybrid Work." Gallup, Inc. Accessed January 2, 2026. https://www.gallup.com/401384/indicator-hybrid-work.aspx.

Gan, Nectar, and David Culver. "China Is Fighting the Coronavirus with a Digital QR Code. Here's How It Works." *CNN*, April 16, 2020. https://www.cnn.com/2020/04/15/asia/china-coronavirus-qr-code-intl-hnk/index.html.

Gibson, William. *Neuromancer*. New York: Ace Books, 1984.

Gibson, William, David Brin, and Anne Simon. "The Science in Science Fiction." Interview by Brooke Gladstone. *Talk of the Nation*, NPR. November 30, 1999. Radio broadcast, MP3 copy, 50:23. https://www.npr.org/2018/10/22/1067220/the-science-in-science-fiction.

Gichuki, Lydia. "AI Goes to School: The Global AI Education Race, Opportunities and Perils." DevelopmentAid, April 30, 2025. https://www.developmentaid.org/news-stream/post/194647/ai-transforming-education.

Grace, Katja, Julia Fabienne Sandkühler, Harlan Stewart et al. "Thousands of AI Authors on the Future of AI." *Journal of Artificial Intelligence Research* 84 (2025). https://doi.org/10.1613/jair.1.19087.

Haagen-Smit, Arie J. "Chemistry and Physiology of Los Angeles Smog." *Industrial and Engineering Chemistry* 44, no. 6 (1952): 1342–1346. https://web.archive.org/web/20170308044020/https://pdfs.semanticscholar.org/1f18/64424a52955b607781d45f007ed8fd1b8002.pdf.

Hamill, Jasper. "Big Boss Is Watching You: The Growth of Workplace Surveillance Technology." *Stack*, September 12, 2024. https://www.thestack.technology/workplace-digital-surveillance/.

Hamilton, Isobel Asher. "Silicon Valley Insiders Say Facebook, Snapchat, and Twitter Are Using 'Behavioral Cocaine' to Turn People into Addicts." *Business Insider*, July 4, 2018. https://www.businessinsider.com/silicon-valley-insiders-tell-bbc-how-tech-firms-turn-users-into-addicts-2018-7.

Haynes, Trevor. "Dopamine, Smartphones & You: A Battle for Your Time." Harvard University. *Science in the News* (blog), May 1, 2018. https://sites.harvard.edu/sitn/2018/05/01/dopamine-smartphones-battle-time/.

Heath, Chip, and Dan Heath. *Switch: How to Change Things When Change Is Hard.* New York: Crown, 2010.

Heikkilä, Melissa. "Five Things You Need to Know About the EU's New AI Act." *MIT Technology Review*, December 11, 2023. https://www.technologyreview.com/2023/12/11/1084942/five-things-you-need-to-know-about-the-eus-new-ai-act/.

Hern, Alex. "Hacking Risk Leads to Recall of 500,000 Pacemakers Due to Patient Death Fears." *The Guardian*, August 31, 2017. https://www.theguardian.com/technology/2017/aug/31/hacking-risk-recall-pacemakers-patient-death-fears-fda-firmware-update.

Hilbert, Martin, and Priscila López. "The World's Technological Capacity to Store, Communicate, and Compute Information." *Science* 332, no. 6025 (2011): 60–65. https://www.science.org/doi/10.1126/science.1200970.

Hill, Kashmir. "Max Schrems: The Austrian Thorn in Facebook's Side." *Forbes*, February 7, 2012. https://www.forbes.com/sites/kashmirhill/2012/02/07/the-austrian-thorn-in-facebooks-side/.

Honan, Mat. "How Apple and Amazon Security Flaws Led to My Epic Hacking." *Wired*, August 6, 2012. https://www.wired.com/2012/08/apple-amazon-mat-honan-hacking/.

Hoofnagle, Chris Jay, Bart van der Sloot, and Frederik Zuiderveen Borgesius. "The European Union General Data Protection Regulation: What It Is and What It Means." *Information and Communications Technology Law* 28, no. 1 (2019): 65–98. https://doi.org/10.1080/13600834.2019.1573501.

India, Freya, and Jon Haidt. "Algorithms Hijacked My Generation. I Fear for Gen Alpha." *After Babel* (blog), November 10, 2023. https://www.afterbabel.com/p/algorithms-hijacked-my-generation.

International Energy Agency. *Global EV Outlook 2024: Moving Towards Increased Affordability*. Paris: IEA Publications, 2024.

Jensen, David. "PIT and the OTA." Lecture presented at the University of Massachusetts Amherst, Amherst, MA, April 14, 2025.

Johnson, Khari. "IBM Walked Away from Facial Recognition. What About Amazon and Microsoft?" *VentureBeat*, June 10, 2020. https://web.archive.org/web/20250416101000/https://venturebeat.com/ai/ibm-walked-away-from-facial-recognition-what-about-amazon-and-microsoft/ (site discontinued).

Joint Chiefs of Staff. *Cyberspace Operations*. Joint Publication 3–12 (R). Washington, DC: Joint Chiefs of Staff, 2013. https://irp.fas.org/doddir/dod/jp3_12r.pdf.

Kang, Cecilia, and Natasha Singer. "Meta Accused by States of Using Features to Lure Children to Instagram and Facebook." *New York Times*, October 24, 2023. https://www.nytimes.com/2023/10/24/technology/states-lawsuit-children-instagram-facebook.html.

Kennedy, Merrit. "Lead-Laced Water in Flint: A Step-by-Step Look at the Makings of a Crisis." *NPR*, April 20, 2016. https://www.npr.org/sections/thetwo-way/2016/04/20/465545378/lead-laced-water-in-flint-a-step-by-step-look-at-the-makings-of-a-crisis.

Knight First Amendment Institute. "Federal Judge Says Challenge in Zuckerman v. Meta Cannot Go Forward Yet." Press release, November 8, 2024. https://knightcolumbia.org/content/federal-judge-says-challenge-in-zuckerman-v-meta-cannot-go-forward-yet.

Kovacs, Eduard. "Security Firm Discloses Medical Device Flaws as Part of Investment Strategy." *SecurityWeek*, August 26, 2016. https://www.securityweek.com/security-firm-discloses-medical-device-flaws-part-investment-strategy/.

Kubrick, Stanley, dir. *2001: A Space Odyssey*. Metro-Goldwyn-Mayer, 1968.

Kunkle, Gregory C. "New Challenge or the Past Revisited? The Office of Technology Assessment in Historical Context." *Technology in Society* 17, no. 2 (1995): 175–196. https://doi.org/10.1016/0160-791X(95)00007-E.

Kurose, James. "Who Controls the Internet?" Online supplement to chap. 1 of *Computer Networking: A Top-Down Approach,* by James F. Kurose and Keith W. Ross. 8th ed. Hoboken, NJ: Pearson, 2020. Posted February 14, 2022. YouTube video, 21:51. https://www.youtube.com/watch?v=xrd4hD_9fS8.

Kurose, James F., and Keith W. Ross. *Computer Networking: A Top-Down Approach.* 8th ed. New York: Pearson, 2020.

Labzovskii, Lev D., Su-Jong Jeong, and Nicholas C. Parazoo. "Working Towards Confident Spaceborne Monitoring of Carbon Emissions from Cities Using Orbiting Carbon Observatory-2." *Remote Sensing of Environment* 233 (November 2019). https://doi.org/10.1016/j.rse.2019.111359.

Levinson, Kaitlyn. "Why California Is Revamping Its Benefits Application Websites." Route Fifty, December 20, 2024. https://www.route-fifty.com/customer-experience/2024/12/why-california-revamping-its-benefits-application-websites/401823/.

Litman, Todd. *Autonomous Vehicle Implementation Predictions: Implications for Transport Planning.* Victoria, BC: Victoria Transport Policy Institute, 2025. https://www.vtpi.org/avip.pdf.

MacMillan, Douglas. "Data Brokers Are Selling Your Secrets. How States Are Trying to Stop Them." *Washington Post,* June 24, 2019. https://www.washingtonpost.com/business/2019/06/24/data-brokers-are-getting-rich-by-selling-your-secrets-how-states-are-trying-stop-them/.

MacMillan, Douglas, David Ovalle, and Aaron Schaffer. "Arrested by AI: Police Ignore Standards After Facial Recognition Matches." *Washington Post,* January 13, 2025. https://www.washingtonpost.com/business/interactive/2025/police-artificial-intelligence-facial-recognition/.

Majewski, Taylor. "It's Time to Retire the Term 'User.'" *MIT Technology Review,* April 19, 2024. https://www.technologyreview.com/2024/04/19/1090872/ai-users-people-terms/.

Marr, Bernard. "How Generative AI Will Change the Jobs of Architects and Civil Engineers." *Forbes,* March 15, 2024. https://www.forbes.com/sites/bernardmarr/2024/03/15/how-generative-ai-will-change-the-jobs-of-architects-and-civil-engineers/.

McBride, Ashley. "Oakland Made Gains in Closing the Digital Divide During COVID. The Classroom Tech Could Be Here to Stay." *Oaklandside,* July 20, 2021.

https://oaklandside.org/2021/07/20/oakland-made-gains-in-closing-the-digital-divide-during-covid-the-classroom-tech-could-be-here-to-stay/.

McCabe, David. "Google Search Results Vary Widely by User, New Study Says." *Axios*, December 4, 2018. https://www.axios.com/2018/12/04/google-search-results-vary-widely-by-user-new-study-says.

McCracken, Harry. "1995: The Year Everything Changed." *Fast Company*, December 30, 2015. https://www.fastcompany.com/3053055/1995-the-year-everything-changed.

Metz, Cade. "Silicon Valley's Elusive Fantasy of a Computer as Smart as You." *New York Times*, May 16, 2025. https://www.nytimes.com/2025/05/16/technology/what-is-agi.html.

Miller, Jen A. "Fern Hollow Collapse Illustrates Poor US Bridge Maintenance." *Construction Dive*, July 5, 2023. https://www.constructiondive.com/news/fern-hollow-collapse-illustrates-lack-of-bridge-maintenance-in-us/654194/.

Miller, Sarah. "The Addictiveness of Social Media: How Teens Get Hooked." Jefferson Health, June 2, 2022. https://www.jeffersonhealth.org/your-health/living-well/the-addictiveness-of-social-media-how-teens-get-hooked.

Mozur, Paul, Raymond Zhong, and Aaron Krolik. "In Coronavirus Fight, China Gives Citizens a Color Code, With Red Flags." *New York Times*, March 1, 2020. https://www.nytimes.com/2020/03/01/business/china-coronavirus-surveillance.html.

Mukherjee, Siddhartha. "Cancer's Invasion Equation." *New Yorker*, September 4, 2017. https://www.newyorker.com/magazine/2017/09/11/cancers-invasion-equation.

Nader, Ralph. *Unsafe at Any Speed: The Designed-In Dangers of the American Automobile*. New York: Grossman, 1965.

National Archives. "Medicare and Medicaid Act (1965)." Milestone Documents. Last reviewed February 8, 2022. https://www.archives.gov/milestone-documents/medicare-and-medicaid-act.

National Institute of Standards and Technology. "NIST Study Evaluates Effects of Race, Age, Sex on Face Recognition Software." Press release, December 19, 2019. https://www.nist.gov/news-events/news/2019/12/nist-study-evaluates-effects-race-age-sex-face-recognition-software.

NEA Task Force on Artificial Intelligence in Education. *Report of the NEA Task Force on Artificial Intelligence in Education*. Washington, DC: National Education Association, 2024. https://www.nea.org/sites/default/files/2024-10/nea-ai-task-force-report-2024.pdf.

Neely, Stephen, Christina Eldredge, and Ron Sanders. "Health Information Seeking Behaviors on Social Media During the COVID-19 Pandemic Among American Social Networking Site Users: Survey Study." *Journal of Medical Internet Research* 23, no. 6 (2021). https://doi.org/10.2196/29802.

New Mexico Department of Justice. "New Mexico Attorney General Raúl Torrez Announces Arrests of Suspected Online Predators Following 'Operation MetaPhile.'" Press release, May 8, 2024. https://nmdoj.gov/press-release/new-mexico-attorney-general-raul-torrez-announces-arrests-of-suspected-online-predators-following-operation-metaphile/.

Nieto-Rodriguez, Antonio. "It's Time to End the Battle Between Waterfall and Agile." *Harvard Business Review*, October 10, 2023. https://hbr.org/2023/10/its-time-to-end-the-battle-between-waterfall-and-agile.

Nix, Alexander. "The Power of Big Data and Psychographics." Presentation at the 2016 Concordia Annual Summit, New York, NY, September 19, 2016. Posted September 27, 2016. YouTube video, 11:00. https://www.youtube.com/watch?v=n8Dd5aVXLCc.

Norman, Donald A. *The Psychology of Everyday Things*. New York: Basic Books, 1988.

Norman, Donald A. "Words Matter. Talk About People: Not Customers, Not Consumers, Not Users." JND.org, November 17, 2008. https://jnd.org/words-matter-talk-about-people-not-customers-not-consumers-not-users/.

NOYB. "Fines Resulting from NOYB Litigation." Accessed September 7, 2025. https://noyb.eu/en/fines-resulting-noyb-litigation.

Office of Science and Technology Policy. *Winning the Race: America's AI Action Plan*. Executive Office of the President of the United States, July 2025. https://www.whitehouse.gov/wp-content/uploads/2025/07/Americas-AI-Action-Plan.pdf.

Oliver, John, host. *Last Week Tonight*. Season 1, episode 5, "Net Neutrality." Directed by John Perota. Written by Ken Avery, Tim Carvell, and Dan Gurewich. Aired June 1, 2014, on HBO. Video, 13:17, June 2, 2014. https://www.youtube.com/watch?v=fpbOEoRrHyU.

Oliver, John, host. *Last Week Tonight*. Season 4, episode 11, "Net Neutrality II." Directed by Paul Pennolino. Written by Jeff Maurer, John Oliver, Jill Twiss, and

Juli Weiner. Aired May 7, 2017, on HBO. Video, 19:32, May 8, 2017. https://www.youtube.com/watch?v=92vuuZt7wak.

O'Neil, Cathy. *Weapons of Math Destruction: How Big Data Increases Inequality and Threatens Democracy*. New York: Crown, 2016.

Organisation for Economic Co-operation and Development. "Artificial Intelligence." Accessed January 27, 2026. https://www.oecd.org/en/topics/policy-issues/artificial-intelligence.html.

Pahlka, Jennifer. *Recoding America: Why Government Is Failing in the Digital Age and How We Can Do Better*. New York: Metropolitan Books, 2023.

Park, Joon Sung, Michael S. Bernstein, Robin N. Brewer, Ece Kamar, and Meredith Ringel Morris. "Understanding the Representation and Representativeness of Age in AI Data Sets." *AIES '21: Proceedings of the 2021 AAAI/ACM Conference on AI Ethics and Society* (2021): 834–842. https://doi.org/10.1145/3461702.3462590.

Pasquale, Frank. *New Laws of Robotics: Defending Human Expertise in the Age of AI*. Cambridge, MA: Belknap Press, 2020.

Peace Corps. *Budget Request to the Office of Management and Budget: Fiscal Year 1995*. Washington, DC: Peace Corps, 1995. https://files.peacecorps.gov/manuals/cbj/1995.pdf?_gl=1*9yola2*_gcl_au*NTE3NjQ5NzU5LjE3NzI4MDQxMDI.*_ga*MTI1MzE4MjU3My4xNzcyODA0MTAy*_ga_7WoV5D49XM*czE3NzI4MDQxMDIkbzEkZzEkdDE3NzI4MDQ0ODkkajM1JGwwJGgw.

Pertwee, Ed, Clarissa Simas, and Heidi J. Larson. "An Epidemic of Uncertainty: Rumors, Conspiracy Theories and Vaccine Hesitancy." *Nature Medicine* 28, no. 3 (2022): 456–459. https://doi.org/10.1038/s41591-022-01728-z.

Playton, Aisha. "Hip-Hop's Copyright Code: A Deep Dive into How the Genre Has Forged Its Own Ethic in IP." De Beer Attorneys. *IP, Decoded* (blog), updated October 27, 2024. https://www.debeerattorneys.com/post/hip-hop-copyright-ip.

Project Management Institute. *Pulse of the Profession 2024: The Future of Project Work; Moving Past Office-Centric Models*. PMI, 2023. https://www.pmi.org/-/media/pmi/documents/public/pdf/learning/thought-leadership/pmi-pulse-of-the-profession-2024-report.pdf.

Prothero, Arianna, Lauraine Langreo, and Alyson Klein. "Which States Ban or Restrict Cellphones in Schools?" *Education Week*, updated July 17, 2025. https://

www.edweek.org/technology/which-states-ban-or-restrict-cellphones-in-schools/2024/06.

Public Interest Technology University Network. "Our Work." Accessed September 7, 2025. https://pit-un.org/about-us/.

Ramasastry, Anita. "Websites That Charge Different Customers Different Prices: Is Their 'Price Customization' Illegal? Should It Be?" *FindLaw* (blog), June 20, 2005. https://supreme.findlaw.com/legal-commentary/websites-that-charge-different-customers-different-prices.html.

Rogers, Katie, Christine Hauser, Alan Yuhas, and Maggie Haberman. "Trump's Suggestion That Disinfectants Could Be Used to Treat Coronavirus Prompts Aggressive Pushback." *New York Times*, April 24, 2020. https://www.nytimes.com/2020/04/24/us/politics/trump-inject-disinfectant-bleach-coronavirus.html.

Roose, Kevin. "A Conversation with Bing's Chatbot Left Me Deeply Unsettled." *New York Times*, February 16, 2023. https://www.nytimes.com/2023/02/16/technology/bing-chatbot-microsoft-chatgpt.html.

Rosenblatt, Kalhan. "ChatGPT Passes MBA Exam Given by a Wharton Professor." *NBC News*, January 23, 2023. https://www.nbcnews.com/tech/tech-news/chatgpt-passes-mba-exam-wharton-professor-rcna67036.

Rothwell, Jonathan. "Teens Spend Average of 4.8 Hours on Social Media Per Day." *Gallup News*, October 13, 2023. https://news.gallup.com/poll/512576/teens-spend-average-hours-social-media-per-day.aspx.

Rupert, Andrea, Shulamith Lala Straussner, Bente Weimand, and Darryl Maybery. "It Takes a Village to Raise a Child: Understanding and Expanding the Concept of the 'Village.'" *Frontiers in Public Health* 10 (2022): 756066. https://doi.org/10.3389/fpubh.2022.756066.

Russell, Stuart. "Beyond ChatGPT: Stuart Russell on the Risks and Rewards of A.I." Interview by Jerry Kaplan. Commonwealth Club of California, San Francisco, CA, April 3, 2023. YouTube video, 1:13:24. https://www.youtube.com/watch?v=ow3XrwTmFA8.

Russell, Stuart, and Peter Norvig. *Artificial Intelligence: A Modern Approach*. 4th ed. Global ed. Hoboken, NJ: Pearson, 2021.

Salman, Javeria. "How One City Closed the Digital Divide for Nearly All Its Students." *Hechinger Report*, April 14, 2022. https://hechingerreport.org/how-one-city-closed-the-digital-divide-for-nearly-all-its-students/.

Satariano, Adam, and Jenny Gross. "Google and Apple Face Billions in Penalties After Losing E.U. Appeals." *New York Times*, September 10, 2024. https://www.nytimes.com/2024/09/10/technology/european-union-apple-google-antitrust.html.

Scally, Derek. "Max Schrems Files First Cases Under GDPR Against Facebook and Google." *Irish Times*, May 25, 2018. https://www.irishtimes.com/business/technology/max-schrems-files-first-cases-under-gdpr-against-facebook-and-google-1.3508177.

Sherman, Justin. "How Shady Companies Guess Your Religion, Sexual Orientation, and Mental Health." *Slate*, April 26, 2023. https://slate.com/technology/2023/04/data-broker-inference-privacy-legislation.html.

Simons, Josh. *Algorithms for the People: Democracy in the Age of AI*. Princeton, NJ: Princeton University Press, 2023.

Singer, Natasha. "How Mark Zuckerberg's Meta Failed Children on Safety, States Say." *New York Times*, June 22, 2024. https://www.nytimes.com/2024/06/22/technology/zuckerberg-instagram-child-safety-lawsuits.html.

Smith, Andrew P., and Hasah Alheneidi. "The Internet and Loneliness." *AMA Journal of Ethics* 25, no. 11 (2023): 833–838. https://doi.org/10.1001/amajethics.2023.833.

Smuha, Nathalie A. "Concluding Remarks." In *The Cambridge Handbook of the Law, Ethics and Policy of Artificial Intelligence*, edited by Nathalie A. Smuha, 429–432. Cambridge: Cambridge University Press, 2025.

Song, Grace, and Zachary Antoyan. "Reflecting on 10 Years of Food Assistance in California." Code for America, August 27, 2024. https://codeforamerica.org/news/reflecting-on-10-years-of-getcalfresh/.

Statista. "Countries with Most Content Available on Netflix Worldwide as of February 2025." Infographic. Statista, February 2025. https://www.statista.com/statistics/1013571/netflix-library-size-worldwide/.

Stewart, Emily. "Lawmakers Seem Confused About What Facebook Does—and How to Fix It." *Vox*, April 10, 2018. https://www.vox.com/policy-and-politics/2018/4/10/17222062/mark-zuckerberg-testimony-graham-facebook-regulations.

Sudhakar, Soumya, Vivienne Sze, and Sertac Karaman. "Data Centers on Wheels: Emissions from Computing Onboard Autonomous Vehicles." *IEEE Micro* 43, no. 1 (2023): 29–39. https://doi.org/10.1109/MM.2022.3219803.

Sullivan, Gail. "What the Heck Is Net Neutrality?" *Washington Post*, May 15, 2014. https://www.washingtonpost.com/news/morning-mix/wp/2014/05/15/everything-you-need-to-know-about-net-neutrality/.

Sutherland, Jeff. "How the Agile Manifesto Came to Be." Posted February 11, 2021. Scrum Inc. YouTube video, 15:06. https://www.youtube.com/watch?v=AGdi3zZutXk.

Thoughtworks. *Responsible Technology Playbook: Tools for the United Nations.* United Nations and Thoughtworks, 2023. https://www.thoughtworks.com/en-us/insights/e-books/responsible-tech-playbook-un.

Tocqueville, Alexis de. *Journeys to England and Ireland (1835)*. New Haven, CT: Yale University Press, 1958. https://sites.pitt.edu/~syd/toq.html#:~:text=104%2D7.,and%20over%20the%20whole%20world.

UN System Chief Executives Board for Coordination. "Principles for the Ethical Use of Artificial Intelligence in the United Nations System." CEB/2022/2/Add.1 (October 27, 2022). https://unsceb.org/sites/default/files/2023-03/CEB_2022_2_Add.1%20%28AI%20ethics%20principles%29.pdf.

US Department of Agriculture. "Recalls and Public Health Alerts." Accessed September 4, 2025. https://www.fsis.usda.gov/recalls.

US Environmental Protection Agency. *Clean Water Rule Comment Compendium: Mass Mailing Campaigns.* Response to Comments for the Clean Water Rule. June 2015. https://www.epa.gov/sites/default/files/2015-06/documents/cwr_response_to_comments_mass_mailing_campaigns.pdf.

US Government Accountability Office. "Closing the Digital Divide for the Millions of Americans Without Broadband." *WatchBlog*, February 1, 2023. https://www.gao.gov/blog/closing-digital-divide-millions-americans-without-broadband.

US Government Accountability Office. "'Why Do I Feel like Somebody's Watching Me?' Workplace Surveillance Can Impact More than Just Productivity." *WatchBlog*, October 29, 2024. https://www.gao.gov/blog/why-do-i-feel-somebodys-watching-me-workplace-surveillance-can-impact-more-just-productivity.

Van Quathem, Kristof. "European Commission Confirms Plans to Simplify GDPR." Covington. *Inside Privacy* (blog), March 17, 2025. https://www.insideprivacy.com/gdpr/european-commission-confirms-plans-to-simplify-gdpr/.

Vaughan, Emma Henderson. "NCMEC Releases New Sextortion Data." *National Center for Missing and Exploited Children* (blog), April 15, 2024. https://www.missingkids.org/blog/2024/ncmec-releases-new-sextortion-data.

Vinsel, Lee, and Andrew L. Russell. *The Innovation Delusion: How Our Obsession with the New Has Disrupted the Work That Matters Most.* New York: Crown Currency, 2020.

Vogels, Emily A. *The State of Online Harassment.* Pew Research Center, January 2021. https://www.pewresearch.org/internet/wp-content/uploads/sites/9/2021/01/PI_2021.01.13_Online-Harassment_FINAL-1.pdf.

Vogels, Emily A., Risa Gelles-Watnick, and Navid Massarat. *Teens, Social Media and Technology 2022.* Pew Research Center, August 10, 2022. https://www.pewresearch.org/internet/2022/08/10/teens-social-media-and-technology-2022/.

Wadhwa, Vivek. "Code for America: An Elegant Solution for Government IT Problems." *Washington Post,* December 19, 2011. https://www.washingtonpost.com/national/on-innovations/code-for-america-an-elegant-solution-for-government-it-problems/2011/12/16/gIQAXrIu2O_story.html.

Walker, Tim. "Take Cellphones Out of the Classroom, Educators Say." *NEA Today,* October 3, 2024. https://www.nea.org/nea-today/all-news-articles/take-cellphones-out-classroom-educators-say.

Wallace, Gregory. "That Was No Cyberattack on the FCC, Inspector General Says—Just John Oliver Fans." *CNN,* August 8, 2018. https://www.cnn.com/2018/08/08/politics/fcc-website-crash-ig-john-oliver-show-hbo-net-neutrality/index.html.

White House. "Blueprint for an AI Bill of Rights." Office of Science and Technology Policy. Accessed September 5, 2025. https://bidenwhitehouse.archives.gov/ostp/ai-bill-of-rights/.

WHO–China Joint Mission on Coronavirus Disease 2019. *Report of the WHO–China Joint Mission on Coronavirus Disease 2019 (COVID-19).* World Health Organization, February 28, 2020. https://www.who.int/docs/default-source/coronaviruse/who-china-joint-mission-on-Covid-19-final-report.pdf.

Who Is Selling Your Data: A Critical Examination of the Role of Data Brokers in the Digital Economy: Hearing Before the Subcomm. on Oversight and Investigations of the H. Comm. on Energy and Commerce. 118 Cong. (2023). Testimony of Justin Sherman, Senior Fellow and Research Lead, Data Brokerage Project, Duke University Sanford School of Public Policy.

Wolford, Ben. "What Is GDPR, the EU's New Data Protection Law?" GDPR.eu. Accessed September 5, 2025. https://gdpr.eu/what-is-gdpr/.

Wong, Matteo. "The GPT Era Is Already Ending." *The Atlantic*, December 6, 2024. https://www.theatlantic.com/technology/archive/2024/12/openai-o1-reasoning-models/680906/.

World Economic Forum. *Future of Jobs Report 2025*. Geneva: World Economic Forum, January 2025. https://reports.weforum.org/docs/WEF_Future_of_Jobs_Report_2025.pdf.

Wu, Tim. *The Master Switch: The Rise and Fall of Information Empires*. New York: Knopf, 2010.

Zuboff, Shoshana. *The Age of Surveillance Capitalism: The Fight for a Human Future at the New Frontier of Power*. London: Profile Books, 2019.

INDEX

Publisher contact:
The MIT Press
Massachusetts Institute of Technology
77 Massachusetts Avenue, Cambridge, MA 02139
mitpress.mit.edu

EU Authorised Representative:
Easy Access System Europe, Mustamäe tee 50,
10621 Tallinn, Estonia
gpsr.requests@easproject.com

Printed by Integrated Books International,
United States of America